ALCHEMY IN THE RAIN FOREST

POLITICS, ECOLOGY, AND RESILIENCE IN A

NEW GUINEA MINING AREA

Jerry K. Jacka

Duke University Press   Durham and London   2015

© 2015 Duke University Press
All rights reserved
Printed in the United States of America on acid-free paper ∞
Designed by Natalie F. Smith
Typeset in Caecilia and Meta Pro by Graphic Composition, Inc.

Library of Congress Cataloging-in-Publication Data
Jacka, Jerry K., [date] author.
Alchemy in the rain forest : politics, ecology, and resilience in a New
Guinea mining area / Jerry K. Jacka.
pages cm—(New ecologies for the twenty-first century)
Includes bibliographical references and index.
ISBN 978-0-8223-5979-1 (hardcover : alk. paper)
ISBN 978-0-8223-6011-7 (pbk. : alk. paper)
ISBN 978-0-8223-7501-2 (e-book)
1. Porgera (Papua New Guinea)—Social conditions—21st century.
2. Porgera (Papua New Guinea)—Environmental conditions—21st
century. 3. Ethnology—Papua New Guinea—Porgera. 4. Gold mines and
mining—Social aspects—Papua New Guinea—Porgera. 5. Gold mines
and mining—Environmental aspects—Papua New Guinea—Porgera.
I. Title. II. Series: New ecologies for the twenty-first century.
DU740.9.P66J33 2015
995.6'305—dc23   2015015086

Cover art: Local boys take in views of the dump sites of the Porgera
Joint Venture Gold Mine from an old mining truck tire, Porgera, Papua
New Guinea, November 17, 2010. Photograph by Brent Stirton / Getty
Images Reportage.

FOR JO

CONTENTS

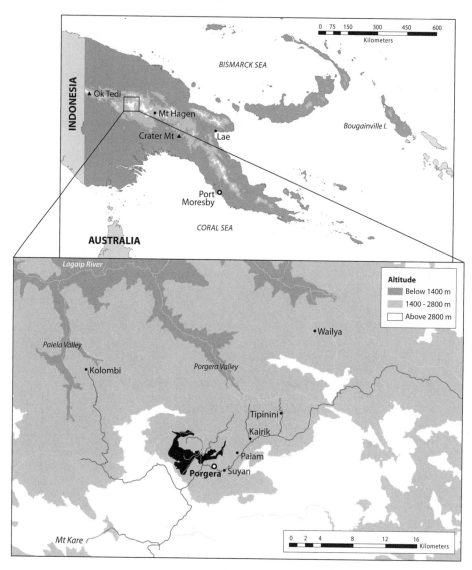

Map 0.1. Inset map of the Porgera valley in Papua New Guinea.

ACKNOWLEDGMENTS

There are a number of people and institutions that I am deeply grateful to for providing me the opportunities to conduct research in Papua New Guinea. The Wenner-Gren Foundation for Anthropological Research (Gr. 6389) and the Porgera Development Authority funded the first phase of my research (December 1998 to February 2000). The second period of research (November to December 2006) was funded by a Faculty Development Grant from North Carolina State University's College of Humanities and Social Sciences. The National Research Institute granted in-country research permission. I thank all of these institutions for their support. All maps were created by me in ArcGIS; any errors are my responsibility. Unless noted, all figures in the captions are also my creation, and all photos were taken in Porgera by me.

In many ways this book was written in a dialogue with an anthropologist whom I never had the privilege of meeting: Roy Rappaport. Rappaport was one of the advisors of my PhD advisor, Aletta Biersack. The introduction to Rappaport's work through Aletta often made me wonder what an updated (third) edition of *Pigs for the Ancestors* would look like with an attention to globalization, Christianity, and colonialism. I hope that in some small measure this book can contribute to such an envisioning.

I am particularly grateful to Aletta Biersack for her support. Her keen intellect and razor-sharp critique have forced me to critically examine my own scholarship and emulate her writings on Ipili speakers. I also had the chance to visit Aletta in her field site in Paiela in 1999 and am deeply appreciative of the opportunity she provided me to meet the people she has been working with for over forty years.

I am lucky to have a cohort of researchers working in Porgera and elsewhere in PNG and the Pacific with whom to discuss ideas, share experiences, present research findings, and circulate drafts of articles and other publications. My interactions with two fellow Porgeran researchers, Glenn Banks and Alex Golub, have been academically productive and have greatly expanded my knowledge of issues and circumstances surrounding the Special Mining Lease area. With Alex I especially enjoyed having another *wantok* in the valley to chat with while pursuing our respective dissertation research topics.

Other PNG and Pacific researchers I would like to thank are Joshua Bell, Jamon Halvaksz, Edvard Hviding, Stuart Kirsch, David Lipset, John Wagner, Holly Wardlow, and Paige West. Jamon's office, just down the hall, has served as a refuge and a place to laugh and share stories about PNG, particularly during those trying times when I felt as if I couldn't possibly keep working on this book. (Thanks also to Jamon and my colleague Thad Bartlett for the beer and pizza lunches during these same trying times.) I am deeply thankful for the support and friendship of Paige West over the past decade and a half; her intellectually stimulating work and her comments on my research over the years have helped shape a number of the ideas in this book. She is the paragon of what a colleague in academia should be.

My colleagues Tom Struhsaker, Matt Lauer, Martha Macintyre, Susan Crate, Anthony Oliver-Smith, Deborah Gewertz, Frederick Errington, Bruce Knauft, Sandra Bamford, Joel Robbins, Bill Clifford, Bob Moxley, Katherine Verdery, Andrew Strathern, Mike Cepek, Phil Shearman, Rick Stepp, Heather Lazarus, and Colin Filer were all supportive. Archival research was conducted at the Melanesian Archives at the University of California–San Diego, National Library of Australia, National Archives of Papua New Guinea, University of Papua New Guinea Library, and the National Research Institute Library (PNG). My deepest thanks to the people who helped with these resources, especially Kathy Creely at UCSD. Thanks also to the graduate students in my Seminar in Resource Frontiers and my various classes on environmental anthropology who aided immensely to the theoretical development of this book. Conversations with my student Jason Roberts have been particularly enjoyable in this regard.

In PNG various people within Porgera Joint Venture provided occasional support, for which I am grateful. Greg McNee in community relations allowed me the use of a phone from time to time and also provided a flight to Mt. Hagen in 1999. Kai Lavu in environmental affairs provided rainfall and temperature data. Ila Temu in management provided a flight from Mt. Hagen to Porgera in 2006.

It has been a pleasure to work with Gisela Fosado, Lorien Olive, and Christine Riggio at Duke University Press. I also thank the two anonymous reviewers for the substantial improvements that came from their comments and suggestions.

I owe a particular debt of gratitude to Edgar Rice Burroughs, whose books, which I read as a child, made me long to see faraway places; and to my parents, Jim and Terry Jacka, who instilled in me a passion for travel and a love of the natural world.

My wife, Joanna Lambert, has been my biggest supporter, patiently suffering the inevitable irascibility that comes with writing a book. Her love and encouragement through this long process have sustained me beyond words. Long walks through the Texas Hill Country with her and our dog, Jupiter, provided much-needed clarity and a fresh perspective during bouts of intellectual uncertainty.

I owe everything in this book to the people of Papua New Guinea, particularly the Porgerans. In Port Moresby, Joe Ketan befriended me immediately and introduced me to the incredible hospitality of Papua New Guineans. Upon arriving in Porgera for the first time, standing with all my luggage in the rain, with night coming on, and uncertain of where to go, Jeffrey Puge, then the mining coordinator for the District Administration, graciously took me into his house while negotiating my entry into the Porgera valley. Jeffrey's continuing friendship during the fourteen months of my dissertation research was most appreciated. Kurubu Ipara, district administrator of Porgera, arranged for my long-term housing and, whenever possible, let me use his office phone to keep in touch with loved ones back in the United States. Kaipas Wangi opened his home to me in 2006; to him and his family I give thanks.

The number of people I am indebted to in Tipinini (eastern Porgera) is too vast to list, but I must single out a few. Peter Muyu, Kipan Wuambo, and Epe Des were my constant companions in 1998–2000. Hardly a day has gone by since leaving them that I have not reflected on my times with them and all they taught me of their culture. Ecological surveys conducted in 2006 could not have been done without the assistance of Epe Des, Jerry Loa, and Solomon and Wanpis Kaipas. Pes Bope, Muyu

Yakati, Tisapa Yawe, and Kainu Muyaka patiently sat with me for hours upon hours, becoming para-anthropologists of a sort as they learned to anticipate my questions about painstakingly detailed minutiae of kinship, cosmology, and the historical past. Finally, Ben Penale became my best friend and a brother. His interest in and support of my research are manifested on every page of this book. This would not be half the book it is without you, *amene*.

Introduction

[handwritten marginal notes: resilience - springs back onto shape / elasticity / recover quickly / ontology - study of nature of being / existence]

This book is an ethnographic, historical, and ecological account of the social and environmental impacts of large-scale mining development in the Papua New Guinea (PNG) highlands. In the chapters that follow, I examine two intersecting themes: first, the making of a "resource frontier" (Tsing 2005) in the homelands of the Porgeran people—predominantly subsistence horticulturalists who are also hosts to the world-class Porgera Gold Mine—and second, how a people whose lives are tied so closely to the land negotiate massive social and environmental transformations. Woven within these two themes are two topics central to environmental anthropology: resilience and ontology. Resilience, or "the ability of . . . systems to absorb changes . . . and still persist" (Holling 1973: 17), compels us to study social and ecological changes not as aberrant but as a means to highlight both the adaptive creativity and the vulnerabilities of coupled social-ecological systems. A focus on ontology, or the study of the nature of being, allows us to examine the various ways that the world is actualized by different peoples (Descola 2013; Viveiros de Castro 1998, 2004) and the implications of the encounters between locally derived ontologies and the ontologies of capitalism and Christianity.

As a resource frontier, the Porgera case is particularly compelling because it complicates simple narratives regarding multinational development practitioners and indigenous people: quite simply, Porgerans complain that there is in fact not *enough* development occurring in their valley (see also West 2012). The Porgera Valley has been the site of alluvial gold mining since the late 1940s, when Australian prospectors paid Porgeran workers with shells and steel tools, then, beginning in the late 1950s, with Porgerans as alluvial miners selling their gold for cash. In 1990 a world-class gold mine opened that was operated by the Porgera Joint Venture (PJV), a consortium of multinational mining companies, the Papua New Guinean state, and customary landowners within the Special Mining Lease area. Porgera is one of only four mines in the world where riverine tailings disposal is practiced (Vogt 2012), and both tailings and waste rock are dumped into the Porgera River. As the hard-rock mining wastes buried the alluvial gold beds, gold mining—which in the alluvial mining years had been a "second garden" (Pacific Agribusiness 1987) for Porgerans—in this new era radically reshaped local and regional political dynamics, processes of social organization and land tenure reckoning, and land management practices. More generally large-scale resource development in PNG has created a new category of citizenry—"landowners" (Ernst 1999; Filer 1997; Gilberthorpe 2007; Golub 2007; Jorgensen 2001)— who, as the recipients of development proceeds and practices, often benefit at the expense of the "nonlandowners" who are their kin and neighbors. The disparities in wealth between the landowners and the nonlandowners have generated significant social tension as most Porgerans have come to realize that the promises of development are not forthcoming in the ways promised by local elites and state politicians. In response to the failure of development, many young men have turned to "working in the life market," a local term that is used to describe the act of engaging in armed conflict against relatively wealthier social groups in the valley in the hopes of extracting mining money in the form of compensation.

Despite all of the tensions and conflicts that resource development can generate, this is not a simple morality tale of evil multinationals dispossessing naïve indigenous rain forest peoples of their land, such as the story depicted in the highest-grossing movie of all time, *Avatar* (2009). Instead the Porgera case highlights the complex and uneven ways in which development occurs and the unexpected outcomes it creates (Smith 2008). While people desire the benefits associated with development (Connell 1997; Errington and Gewertz 2004; Gewertz and Errington 1999) and actively court resource exploration on their own lands, they simultaneously describe these changes as undermining the very foundations of their universe (see also Kirsch 2001). For Porgerans gold is a cosmological resource in that gold deposits are the products of ancestral spirits; in the case of the mine, gold is the shed skin of a python spirit that their ancestors made sacrifices to in the recent past. Other groups in the area also sacrificed to python spirits, and consequently gold is believed to permeate the entire region. The sacrifices that were fed to the python spirits were the pigs and garden produce that came from human labor. In exchange for these gifts, the spirits ensured that the land remained fertile and that people, crops, and animals were healthy and fecund. The reciprocity that existed among people, the land and its resources, and the spirits has withered with the advent of large-scale mining, conversion to Christianity, the recent importation of firearms, and limited integration into capitalist markets. This demise of reciprocity is captured in the phrase *yu koyo peya*, "the land is ending," which many Porgerans see in the transformations of various ecological and social processes in their lives.

To address how the land is ending, I examine social memory (Cruikshank 2005) and the forms of knowledge and morality that humans use to engage with the environment from a Porgeran perspective (Viveiros de Castro 1998, 2004) and the implications that these moral and cognitive systems have for practical engagement with the land and its resources. In Porgera landslides, earthquakes, and rivers that disappear into limestone sinkholes and flow underground all highlight the idea that the landscape is active and alive and thereby open to ritual and practical manipulation by humans. Spiritual entities that live in the ground provide omens of future events that Porgerans read as having tangible effects on their society. Good relations with these nonhuman spirits are said to result in the health and productivity of people, plants, and animals. In essence, as one researcher described for the neighboring Huli people, there is a "sense of the interconnectedness of geological and human affairs" (Frankel 1986: 17). I would go one step further for the case of Porgera and argue that there is an interconnectedness of the total environment—geology, ecology, climate, and so on—and human affairs. This book explores this total environment and the contradictory yet mutually reinforcing, tensions between the active quest for socioeconomic development and the ending of the land that such development brings.

Most important, this study is not a lament of change, what Marshall Sahlins (1999: iii), borrowing from development discourse, so aptly called "despondency theory." Instead I seek to understand creative and dynamic responses to change and uncertainty as critical elements in theorizing Porgeran peoples' involvement with such diverse processes as horticultural production, multinational mining ventures, climatic variation, and Christian missionization. I draw on insights from both cultural and ecological theory, in which change and transformation are central to social and environmental systems. In the social sciences there are a number of scholars whose works are instrumental in understanding how capitalism and modernization are appropriated and refracted through local cultural schemes and practices (Ferguson 1999, 2006; Ong 2010; Sahlins 1988, to name but a few). From ecology the concept of resilience offers a useful frame of reference for understanding social and environmental resilience (see also Berkes et al. 2003b; Gunderson and Holling 2002). At the same time, though, I temper this with an awareness of the importance of "place" (Basso 1996a; Escobar 2008; Massey 1994), or the enduring associations that people have with their landscapes through their subsistence livelihoods, histories, and myths. Studies of place offer insights into sources of stability, or "groundedness," between people and the lands they inhabit. I also take seriously the effects of development and Christianity on Porgeran ontologies (cf. Cruikshank 2012; Descola 2013; Kohn 2007; Latour 2009; Nadasdy 2007; Viveiros de Castro 1998, 2004), addressing what Eduardo Kohn (2007: 6) refers to as "an anthropology of life," which is "the practice of a kind of anthropology that situates all-too-human worlds within a larger series of processes and relationships that exceed the human." While this new attention to ontology in anthropology allows us to expand the scope of our inquiry to the relationships between humans and nonhuman others, this book also examines the implications of *ontological dissonance*, or the implications of how people's lived-in worlds transform due to external events and influences.

Besides change, I seek to understand struggles over access to and control over land and resources, one of the key concerns of "political ecology" (Biersack 2006b; Blaikie and Brookfield 1987b; Paulson and Gezon 2005; P. Robbins 2004). Political ecology, according to Blaikie and Brookfield (1987a: 17), "combines the concerns of ecology and a broadly defined political economy. Together this encompasses the constantly shifting dialectic between society and land-based resources, and also within classes and groups within society itself. [It] also derive[s] from political economy a concern with the role of the state." Struggles and conflicts over land and resources in Porgera have resulted in large areas of deforestation and forest degradation, another major concern of political ecology (P. Robbins 2004). In this light I attend to Vayda and Walters's (1999: 169) critique that within political ecology what is often studied "are political controls or political contests over natural resources and not, or at least not to any significant extent, how the resources are affected by those controls or contests." Combining ecological sampling with remote sensing, I demonstrate that deforestation is detrimentally impacting areas adjacent to the mine and main roads in Porgera. Defor-estation and forest degradation are complex phenomena, however, and understanding these processes highlights that population pressures and development alone are not enough to explain habitat change. Property rights, kinship connections, spiritual beliefs, habitat types, and rules over resource management all intertwine to produce varying outcomes on forests in Porgera.

While it is essential to ensure that political ecologists attend to ecol-ogy, it is also critical that we define whose ecology we are talking about (West 2005). There is a long tradition in anthropology of studying indig-enous ecological knowledge (Berlin et al. 1973; Conklin 1954; Nazarea 1999). However, research on indigenous ecological systems needs to move beyond the narrow focus on classification and taxonomic categori-zation that defines much of this work. As Hviding (1996: 26) argues, "Sys-tems of classification must be considered in a wider context, comprising cognition, productive activities, and social relations." To this end I devote a major portion of this book to detailing an indigenous philosophy of ecology, or an ontology, as Porgerans expressed it to me. A key concept in Porgeran ecological understanding is the flow of a vital force called *ipane* or "grease" that permeates all living things. Correspondingly the flow of grease across trophic levels sustains the social-ecological system. However, in Porgera the food web involves not just the feeding rela-tionships of biotic organisms but also includes spirits and the land (cf. Rappaport 1984). The sacrifice of pigs and the burning and offering of pig fat (*yia ipane*) to ancestral spirits resulted in the spirits making the land fertile—giving it *yu ipane*, "land grease." Plants and trees are dependent on land grease and incorporate it during their growth cycle. The con-sumption of plants and food crops by pigs and humans puts grease into their bodies. The grease of pigs flows to both ancestral spirits through sacrifice and to humans through consumption. The ecological system in Porgera tends toward entropy, however, as the land grease (yu ipane) and the grease in humans (ipane) is expended through production and reproduction, ultimately culminating in a catastrophic destruction of the world so that it can be reborn again. However, it is primarily through human labor (*piape*, "work") that entropy and catastrophe are mitigated.

Porgeran ecology, as such, is not the ecology of equilibrium or ho-meostasis between social and ecological systems, such as depicted in an earlier era of environmental anthropology in the PNG highlands (Rappaport 1984). Geological and climatic anomalies, such as volcanic eruptions, landslides, droughts, and frosts, and social disruptions, such as endemic warfare, Christianity, and multinational resource extraction, have contributed to a suite of dynamic and creative social and environmental practices and perceptions by Porgerans regarding their precarious existence. In fact it is anticipation of uncertainty—crop failures, tribal fighting, spirit attacks—that defines Porgeran socioenvironmental practices. This same attitude permeates their recent involvement with colonialism, gold mining, and Christianity. Yet there are limits to resilience (Adger et al. 2009; Nelson et al. 2007; Ogden et al. 2013), and in the final chapters of this book I examine these in more detail in order to contribute to the growing chorus of voices that seek to promote sustainable livelihoods in the face of ecological degradation and capitalist intrusion.

Alchemy in the Rain Forest

As mentioned, one of the main themes in this book is to document the making of a resource frontier in Porgera, especially from the perspective of nonlandowners who receive few to no benefits from mining. An analysis of a resource frontier highlights the ways that resource development commodifies nature and social relations, the implications of these processes, and the new ways that people think about and act on each other and their environments (Agrawal 2005). The second main theme in this book is about change—cultural and environmental change and how they are inextricably linked—captured by the phrase the land is ending. For a horticultural society to say that the land is ending is more than just a statement of the perceived degradation of the material environment. It is a statement that encompasses transformations in concepts of personhood, social relations, kinship dynamics, and cosmological understandings. Traditionally Porgerans engaged in large-scale ritual efforts to prevent the end of the land. Paradoxically the very forms of change that Porgerans desire through development amplify traditional concerns about the land ending. Environmental destruction brought by mining, Christianity with its apocalyptic end times, and the social devastation wrought by the adoption of firearms entangle concerns about the end of the land with the moral decline of society.

My title, Alchemy in the Rain Forest, draws attention to two sets of processes regarding change that encompass this larger ontological perspective: the transformation of land and the transformation of people. These are two inseparable and interlinked processes that have both material and ideological consequences. The goal of alchemy in the ancient and medieval world was to transmute base metals into gold. I use the imagery of alchemy to explore the implications of turning Porgeran lands into gold, which affects the material, social, and cultural landscapes in the region. In material terms, tailings waste dumps, riverine pollution, and deforestation are detrimental legacies of the mine. At the same time, though, infrastructure development, electrification, and other material benefits are viewed in a positive light by most Porgerans. Socially Porgeran society has been bifurcated between those people who receive monetary and other benefits from mining and those who do not. A key feature of social life in the valley today centers around (often violent) political contestations over institutions related to land tenure and social organization. Alternatively some Porgerans have been able to mobilize their social capital through broadened social networks, giving them access to places and opportunities closed to them just a few decades ago. In a cultural view, gold is said to come from the shed skin of an ancestral python spirit that is linked to regionwide social and ecological renewal rituals. Gold extraction therefore threatens the very cosmological underpinnings of the Porgeran worldview. In contrast to this, some see this gold as a gift from the python spirit that allows them to channel mining wealth into their kinship networks and communities (Golub 2006).

A second, perhaps loftier goal of alchemy sought the transformation of people by improving the human condition (Moran 2005; Newman 2005; Smith 1997). In alchemical science the philosopher's stone, the source of the elixir that would transmute metals into gold, could also be used on the human body to prolong life, bring enlightenment to its discoverer, and refine both body and soul (Linden 2003). In this sense the alchemy of development depends on images of progress, advance, and betterment. Since first contact with the outside world Porgerans have fervently sought to incorporate outsiders and their ideas, first in their efforts to obtain the wealth of outsiders through ritual means, then using imprisonment by colonial agents to learn Tok Pisin and become acculturated in colonial practices as a path of upward mobility. Through mining, Porgerans and non-Porgerans alike have envisioned the transmutation of Porgerans from tribal horticulturalists to entrepreneurs and wage laborers embedded in capitalist processes. From the perspective of Christianity, Porgerans see their involvement in a global religion as ushering in an era of not just economic development but also social transformation. As scholars across the social sciences have shown, however, development is itself highly contested as both a concept and a practice (Escobar 1995; Ferguson 1994; Sachs 1992; Scott 1998). Exploring the pursuit of alchemy in Porgera, then, allows us to examine the motivations and practices of a variety of agents and institutions at various scales across space and time. It also lets us explore the materiality of resources (Bakker and Bridge 2006) as we examine how gold—as an agent that shapes human social relations itself—is simultaneously a cosmological substance, an extractive resource, and a globally sought-after commodity that links Porgerans with other Papua New Guineans, transnational mining corporations, international financiers, and everyday consumers. As such it provides a means for us to understand the everyday politics of resource development in an "out-of-the-way place" (Tsing 1993) and the real-world impacts of humanity's desire for gold on the lives and lands of a people who themselves had no word or use for the mineral until the mid-twentieth century. Moreover the metaphor of alchemy highlights the point that as much as we may want to control and manage social and environmental change, the end results may be just as elusive for us in the twenty-first century as they have been for alchemists in the past.

The Global Demand for Gold

In Porgera I was often asked, "Where does all of this gold go? What do people do with it?" Porgerans had been alluvial-mining it themselves for fifty or so years and of course knew of the wealth that was associated with selling it. But what outsiders were doing with the gold they took from the valley was beyond their imagination, as they have no indigenous use nor word for *gold*. During fieldwork in the late 1990s, the question caught me off guard. I assumed that it was used for jewelry and industrial applications, and maybe global finance, but in what proportions I had never thought to consider.

Surprisingly, most of the global demand for gold is in the form of jewelry and gold bars, primarily used as a form of wealth storage in developing countries. According to the World Gold Council's Global Demand Trends reports (available at www.gold.org), between 2001 and 2010 about 68 percent of the global demand for gold went into the jewelry market. Investment (gold bars, coins, and exchange traded funds) accounted for 20.5 percent of the demand, while technology accounted for 11.5 percent. In 2010 four countries—India, China, Turkey, and Saudi Arabia—accounted for 58 percent of the jewelry, bar, and coin demand in the world. The average annual gold demand from 2001 to 2010 was 3,569 metric tons. The supply for this demand comes from two sources: mining and recycled gold jewelry. In 2010 mines around the world produced 2,543 metric tons of gold, while 1,653 metric tons came from recycled gold. Since opening in 1990 the Porgera Gold Mine has produced between 0.7 and 1.3 percent of the world's gold demand annually (18.4 million ounces from 1990–2013; 766,150 ounces on average yearly)—a fractional amount for such a huge impact on Porgeran lands and lives.

Given the diverse and seemingly contradictory tensions between socioeconomic development and biodiversity conservation, it is imperative that we seek sustainable futures that improve peoples' lives while preserving global biodiversity. This can be an extremely difficult goal to achieve (Oates 1999). However, given that humans dominate nearly all major ecosystems today (Vitousek et al. 1997), it is apparent that any solution will require us to understand problems in conservation and development as aspects of linked social-ecological systems. The concept of resilience is one avenue of inquiry that offers promise for resolving the dilemma between development and conservation (Gunderson and Holling 2002; Leslie and McCabe 2013; Nelson et al. 2007). Coupled social-ecological systems that are resilient are able to absorb perturbations, adapt to changing circumstances, and yet still retain their same function and structure (Berkes et al. 2003b). Papua New Guineans have a forty-five-thousand-year history (Summerhayes et al. 2010) of living in their diverse environments, during which time they have adapted to the end of the last Ice Age, the independent invention of agriculture, the arrival of new crops and domesticated animals, and the benefits and ills of the past century of globalization—truly a case of resilience if ever there were one. Yet over the past decade or so, conflicts over resource development and environmental conservation have resulted in the loss of lives, destruction of property, ecological degradation, and other harbingers of severe social-ecological stress (Ballard and Banks 2003; Connell and Howitt 1991; Filer 1999; Hyndman 1994; Imbun and McGavin 2001; Kirsch 2006; May and Spriggs 1990; West 2006a). All of these pose serious challenges to social and ecological resilience in PNG. A critical concern for environmental anthropology is to examine how human societies respond to changes in social-ecological systems in ways that are both adaptive and maladaptive (Rappaport 1993). By doing so we can explore responses and processes that offer hope for finding sustainable solutions to the challenges that face us in the fields of development and conservation.

This book complements recent works in PNG on resource development (Golub 2014; Kirsch 2006; West 2006a) yet also stakes out new theoretical and methodological ground. Paige West's long-term research among the Gimi, who figure centrally in a conservation and development project at Crater Mountain Wildlife Management Area, highlights the contentious and divergent expectations over development between conservation NGO workers and local communities. These tensions eventuated in the demise of conservation efforts around Crater Mountain as locals aligned themselves with mining companies who promised to bring them the development the Gimi hoped for but which conservation had not delivered in the ways they anticipated. To date the Gimi are still waiting for their mine as the prospect is still in the exploration stage. Working in the communities some distance downriver from the world-class Ok Tedi copper and gold mine, Stuart Kirsch's book examines the environmental costs of mining on a lowland Papua New Guinean society. Like Porgera, the Ok Tedi mine practices riverine tailings disposal that has devastated the aquatic ecology and subsistence of the Yonggom, as well as destroying riverside gardens and forests several kilometers inland from the river banks due to siltation-caused flooding. Similar to this book, Kirsch situates these impacts in an indigenous framework, describing the ontological dissonances that a lowland PNG society faces through the deterioration of their environment. While both West and Kirsch work in a political ecology framework and provide compelling accounts of mining (actual and imagined) on indigenous lifeworlds, neither ethnography is situated among a group of people living directly adjacent to, and hence impacted along multiple dimensions socially and ecologically by, a large-scale mine. One of the methodological contributions this book makes is to situate the social and ontological changes that West and Kirsch also discuss alongside empirical observations of environmental changes through remote sensing and ecological sampling. Alex Golub's recent book on elite landowners in Porgera is an excellent companion to this book. Golub and I overlapped in Porgera for a period of several months in 1999 and 2000, and while I worked among nonlandowners, Golub worked among landowners. His book is a contribution to political anthropology in that he examines how collectives (what he calls "leviathans," in this case Ipili landowners and the mine) form out of the actions of disparate individuals pursuing their own ends. While Golub provides a nuanced microanalysis of negotiations over a proposed resettlement scheme, this book explores the ontological and ecological stakes of resource development from a more holistic perspective.

Research in Porgera

Determining what to call the people of Porgera is difficult. The term that I and other researchers have used is *Ipili*. People like Ben and Peter fit the description of Ipili well (i.e., Ipili-speaking parents, Ipili ancestors, long-term family history in the valley). Accounting for people like my Enga-speaking friend Epe Des (figure I.1), however, is why I use the term *Porgeran* to describe the people in this book rather than *Ipili*.

The ethnographic data used in this book come from two research trips that total sixteen months. When not in Porgera, I followed events from afar through a variety of sources: news articles, emails and letters from Porgerans, conversations and articles from other scholars working in the area, and corporate reports. I have also analyzed environmental changes from ecological surveys in 1999 and 2006 and remotely sensed satellite imagery and aerial photos from 1972 to 2013 to document the growth of the mine, deforestation and land cover changes, and infrastructure expansion. While most of the data in this book come from ethnographic research, the rest derive from these supplementary sources.

From December 1998 to February 2000 I lived in the eastern Porgera valley in the hamlet of Kolatika while conducting research for my dissertation. The entire eastern Porgera valley is called Tipinini by outsiders; it comprises Ipili speakers, the original inhabitants of the Porgera valley, Enga speakers, and some Huli speakers; the latter two groups have mostly married and migrated in since mining began. I interviewed over two hundred people living in approximately ten hamlets throughout Tipinini and also among other communities along the main road, such as Kairik and Paiam. I also interviewed government workers at Porgera Station (the government headquarters) and mining officials at the Porgera Mine. During this first research period I also worked in several hamlets in Wailya, a community of Enga speakers just to the northeast of Tipinini. In addition to holding semistructured and unstructured interviews, I also conducted participant observation, working with, eating with, and accompanying people throughout their daily routines of work, church, and leisure. Throughout 1999 I conducted resource mapping exercises in several hamlets, gathering data on preferred sites for hunting, gathering forest resources, and harvesting wild pandanus.

During this initial research period I became closely integrated into a subgroup of the Tokoyela clan, the landowning group at Kolatika. Kolatika served as a government base in the eastern Porgera valley and was the location of a patrol officer's house and an agricultural extension officer's house. Patrol officers serve as liaisons between the government and community, but the post was currently vacant at Kolatika, so the government allowed me to live in the patrol officer's house during my dissertation research. I shared this house with two research assistants, Peter Muyu (figure I.2) and Ben Penale (figure I.3), both from the Tokoyela-Maia subclan. Ben and Peter were both married, and each had three children ranging in ages from two to nine. Their families, who lived a few minutes' walk away, absorbed me graciously into their lives and homes.

Figure I.1. Epe Des.   Figure I.2. Peter Muyu.

In 1971 Epe Des left Yelum, his home area among the western Enga, which is several days' walk from Porgera to the northeast across the Lagaip River. He had heard that gold mining companies had moved into the Porgera valley and men were needed as laborers. Arriving in Porgera, Epe soon found out that all of the jobs were taken, but he had a cousin in the eastern part of the Porgera valley (Tipinini) and stayed with him a while to see if any jobs would open up. While living there, he met and fell in love with Isam Tiyupe, an Ipili woman. They married and returned to Yelum, but the constant tribal fighting prompted Isam to return to Porgera within a year or so. Epe stayed to fight and help his group with compensations, but after one battle he was shot six times with arrows and nearly died, which convinced him to leave Yelum permanently. As soon as he was well enough to make the journey, he walked back to Porgera to join Isam. They settled in the Tipinini area, maintaining homes simultaneously at various hamlets, such as Lese, Kolatika, and later Kukulama. Over the years they had four children. When I first became acquainted with Epe in 1998, he had recently been elected to his fifth term as the village councilor for Tipinini #1 ward (local-level government districts are divided into census wards, of which there are ten in

Figure I.3. Ben Penale.

Figure I.4. Research assistants in 2006. Jerry Loa, Wanpis Kaipas, Epe Des, and Iki Peter (*back, left to right*); Timoti Peter and Kala Kaipas (*middle, left to right*); Solomon Kaipas (*front*).

the Porgera valley) and was also the Seventh Day Adventist (SDA) pastor at Kolatika.

Epe's case of being an outsider and also a leader is not unique and illustrates the need to think creatively about ethnic identities in this region of the highlands. An immigrant Engan married to an Ipili woman, Epe came to dominate political and religious life for thirty years in the eastern Porgera valley as a pastor and councilor. At times in this book I use categories like Enga, Ipili, and Huli, mostly to talk about language and precolonial ethnic identities, but I also use Porgeran to capture some of the dynamics that have resulted from the long-term influx of people from surrounding communities to this area. As Paul Sillitoe (1999) argues for the nearby Wola of Southern Highlands Province, fixedness of identity and localization of the group is what New Guinea highlanders are working against socially (see also Biersack 1995a). The movement of families and groups across the landscape allows people to continually renegotiate issues of boundary and identity, which is crucial in these acephalous, stateless societies. The intrusions of the Papua New Guinea state and developers who cannot accommodate such social flux are a key source of the tensions in contemporary highlands societies, a point that is central to understanding how introduced vulnerability challenges resilient social-ecological systems.

In November and December 2006 I returned to Porgera excited to see all of my old friends and acquaintances. While I normally traveled by public bus to Porgera from the airport at Mt. Hagen, I had just given a talk at the University of Papua New Guinea, which was attended by one of the upper-level managers of Barrick Gold, PNG—the current operators of the mine—who arranged for me to fly in on a Porgera Joint Venture plane. (A friendly gesture despite the fact that my talk was critical of how restricted development was in Porgera for most Porgerans. This same Barrick executive also dashed my hopes regarding studying mine closure.[1] With the recent uptick in the price of gold and the discovery of some new gold-bearing veins, the earliest that mining would end would be in 2032!) I flew into Porgera as the only passenger on a forty-seat airplane; the remainder of the seats were filled with bundles of the latest edition of the national newspaper, the *Post-Courier*, and cartons upon cartons of eggs. As I left the razor wire–topped, fenced enclosure around the Kairik airstrip, set my bags on the side of the road, and waited for a passing vehicle to take me to Tipinini, the airport security guards struck up a conversation with me. They wanted to know where I was going, and I told them Kolatika. They looked shocked and said, "You can't go there, they're in the middle of a big fight now." I learned that two clans were fighting over a compensation payment from PJV. More disturbing, two women and a man, all Huli from the Tari area, had been killed the day before in the midst of fighting on the road to Kolatika. Even the security guards were disturbed. "Women being killed, can you imagine?" one of them asked me.

Bewildered about where to go, I made my way to Porgera Station to see if I could find a room at Porgera's lone hotel, the Mountain Lodge, a place that evokes none of the qualities that the name promises. Outside the razor-wire fence surrounding the cobbled-together temporary housing facilities that are the lodge, I was overjoyed to see that the security guard was a Tipinini man from Kolatika, Jakob Inama. Jakob was amazed to see me, and we both lamented the fighting that had broken out in Tipinini. I asked him to send word to Epe (mobile phone service was still a few years away) that I was in Porgera and needed a place to stay now that it seemed I couldn't return to Kolatika. He promised he would, and we parted.

The following morning Epe arrived at the Mountain Lodge, and we sat in the grass and gazed down the Porgera River valley and told each other how our lives had transpired during the previous six years. In 2003, the same year I defended my dissertation and took a university position, Epe became the pastor of the SDA church at Kukulama, built a permanent materials house next to the church, and moved his family there. In 2005 the Tipinini fight started; to avoid the violence Epe moved down the road to Kairik. (In April 2008 I received an email from Epe's son, George, who said that the house and church in Kukulama had been burned down during another tribal fight.) Epe's family was doing well despite the disruptions; George was working in community affairs at the mine; his elder daughter, Angie, was married and raising children; his daughter Rose was working for an NGO in the valley that worked on women's issues; and his younger son, David, was in high school. Epe encouraged me to come live in Kairik at a hamlet called Apone Camp. He lived next door to the Lutheran pastor of Kairik, Kaipas Wangi, an Enga man from Maramuni, who had already agreed to let me stay at his house. I knew few people in Kairik, although many of them remembered seeing me during my first stint in Porgera, and so with few options, I reluctantly moved to Apone Camp adjacent to the Kairik airstrip.

Over the following weeks, as word spread throughout the eastern Porgera valley that I was living in Kairik, my friends from Kolatika came to visit. The reunions were bittersweet. While some of my younger friends were now married and starting families, many of my closest middle-aged and elder friends and informants had died of old age or other ailments or, more commonly, been murdered in revenge killings. Kipan, whom I met my first night in Kolatika in 1998 and who thereafter was a constant companion, came hobbling into the Lutheran church compound one afternoon on crutches, suffering from a festering wound in his lower leg. Due to all the fighting, he had made the difficult journey along forest trails to spend an afternoon with me and "see [my] face again." I gave him money for medical treatment, and we spent the afternoon reminiscing about the past and pondering what the future held for us. Within a few days Peter and Ben each came separately to visit. Peter was suffering from malaria that he had picked up while mining for alluvial gold in the lower reaches of the Porgera valley. Ben had also experienced some rough times since 2000; due to the fighting he had lost a government job he had held for five years. Since alcohol had been legalized for sale and consumption in Porgera since 2002, he was now the co-owner of a beer hall in Kukulama; however, in emails in 2013 he writes that he now has a government job again with the Enga Provincial Government.

While I lived at Apone Camp, the fighting between the two clans raged on. Every day we heard gunshots in the distance, and every couple of days someone else was killed. I hired Epe and several young men—Solomon and Wanpis, Kaipas's teenage sons, Kala, their cousin and Timoti and Iki Peter, two young men from the Bipe clan whose land we were living on—to help me do ecological surveys. Jerry Loa, a Bipe man about my own age (fortyish), joined our team after the first day to help with land use histories and to aid in negotiating access to his clan members' gardens and forested lands (figure 1.4). Our days were active and full of camaraderie as we surveyed primary- and secondary-growth forests, measured gardens, and compiled garden inventories. The young guys would sing Enga songs and crack jokes about the differences between "pomberena [black, in Ipili] Jerry" (Loa) and "one [white, in Ipili] Jerry" (me). It was hard to ignore the turmoil that the fighting was causing in people's lives, though. Every morning we would wait to hear where gunfire was coming from before we set out in the opposite direction to begin our work. Most days we would run into people singly and in groups furtively making their way along a forest trail to pass through the war zone en route to Porgera Station. Men armed with spears and bows and arrows would seek us out in the forest to drive us off, thinking we were a raiding party from one of the two clans. Upon finding a group of guys counting and measuring trees and writing down all of their cultural and economic uses, they would relax and hang out with us for a bit before continuing on their patrols.

Eventually the fighting reached Apone Camp. During my second month there, in the middle of the night a young man was shot and killed in the center of the hamlet, less than fifty meters from where I was staying. A few days earlier, as I was watching the funeral procession of a politician who had died (from sorcery, everyone said) in Port Moresby leave the Kairik airstrip, a group of young men wielding machetes sidled up behind me. Flustered, and for the first time in Porgera fearing for my safety, I told them to stand somewhere else. They refused to do so, and only when Epe yelled at them did they move on. With these two recent events I realized it was getting too dangerous to remain in Porgera. The morning after the killing in Apone Camp I caught a bus to Mt. Hagen to begin the long journey home. My last memory of Porgera is seeing Jerry Loa, wearing the Australian bush hat I had given him for a present, waving good-bye to me while standing alongside the road in Kairik in the pouring rain.

In the intervening years I have felt challenged about how to write about my experiences there. Given the complexities of the situation in Porgera, there are a range of viewpoints that one could choose to convey the events and circumstances surrounding the development of mining there. As an anthropologist, my primary perspective aligns with the community, although I underscore the ways this is a perspective fraught with factional differences based on locale, gender, age, and other factors. As an environmental anthropologist, my goal is also to argue for the indivisibility of social from environmental factors. How Porgerans conceptualize the changes they have experienced through engagement with mining is as much a story about cultural as environmental changes and the transformations that structure people's interactions with the land. From this perspective, the environment itself has an agency that helps to shape our understanding of this story. As an environmental scientist trained in geographic information system and ecological survey methods, I focus on the impacts of mining and development on the biophysical environment to understand the relationships between resilience and vulnerability in social-ecological systems. Finally, as a political ecologist, I provide a perspective that examines how capitalism, in different ways at different scales, affects concepts and institutions related to property rights, resources, and environmental conservation. My main goal, though, is to tell a story that doesn't reduce the Porgeran experience to one of complete despair in the face of capitalist development, on the one hand, nor one that portrays the bright, hopeful future as represented in glossy brochures produced by the mine, on the other hand. A better story about Porgera is a story that examines change and the ways that complex social and environmental dynamics both foster and challenge sustainable livelihoods in order to build resilient responses to the problems that face us in the twenty-first century (Ogden et al. 2013).

The Structure of the Book

The ethnographic chapters are divided into three parts: "The Making of a Resource Frontier," understanding "development" in its historical and contemporary contexts through such diverse people and institutions as Porgerans, colonial agents, Christian missionaries, the PNG state, and multinational mining companies; "Indigenous Philosophies of Nature, Culture, and Place," detailing Porgeran interactions with their environment, both materially as a source of subsistence and symbolically as a source of cultural meaning; and "Social-Ecological Perturbations and Human Responses," documenting changes in the coupled human-environment system to explore the adaptive and maladaptive responses people make to large-scale resource development in Porgera.

In chapter 1 I position Porgera within contemporary debates in Papua New Guinea over resource development and concerns about conservation. PNG is one of the most culturally, linguistically, and biologically diverse places on the planet, yet this immense repository of human and nonhuman diversity is threatened by capitalist resource extraction and the changing social relations wrought by capitalism (Durkheim 1997; Marx 1990; Weber 1997). Conceptually I deploy theories of resilience, political ecology, ontology, and place to describe the "different landscapes and ecological dynamics" (Scoones 1999: 494) at play in the making of Porgera's "resource frontier." In chapter 2 I focus on the historical encounters between Porgerans and outsiders. Gold prospectors, colonial patrol officers, and Christian missionaries formed the foundation of Porgeran experiences of the outside world, and gold, guns, and the Bible remain essential elements of Porgeran identity today.

Part II delineates a Porgeran ecological philosophy that is developed in chapters 3 through 5. There is no word for *ecology* or *nature* in Porgera, and as such, beliefs and practices associated with the environment contain "an extraordinary amount of human history" (Williams 1980: 67). In other words, to describe Porgeran ecology it is necessary to discuss kinship relations, social organization, spirit beliefs, and other such "social" factors that in Western tradition are not considered part of "natural" systems (see also Descola 2013). The Porgeran social-ecological system pivots around three critical concepts: land, people, and spirits. The essence that unites these, as mentioned, is grease or ipane, and a reciprocal triad of people, spirits, and the land are key for the maintenance of ipane in the system. Chapter 3 deals with how the land is conceptualized cosmologically and the implications of this for local resource management. In chapter 4 I address people, work, kinship, and social organization, and in chapter 5 I explore spirits and their role in earth fertility rituals and the transformation of spirit beliefs in the present.

In part III, I explore resiliency and vulnerability in the face of social and ecological perturbation. Despite being in the tropics, livelihoods in the high-altitude ecology of Porgera are vulnerable to frosts, droughts, and other environmental calamities. Likewise significant levels of intergroup violence are a constant threat. To counter these challenges, the social and ecological practices that I detail in part II emphasize flexibility and openness to change in order to foster resilience. In chapters 6 and 7 I evaluate this flexibility and openness in more detail but also highlight the unexpected forces and processes that are increasing ecological and social vulnerabilities. In particular in chapter 6 I examine human responses to volcanic eruptions, droughts caused by El Niño–Southern Oscillation, and deforestation. Oral and other historical sources are integrated with remote sensing analysis and forest plot surveys to decipher local and regional environmental changes. In chapter 7 I take a detailed look at social disruptions brought by mining development. Benefits from mining create "wealth without work" (Reed 2009: 3) and reshape attitudes about productive labor. For many of the disenfranchised young men left out of the mining wealth, warfare acquires new value as a means to extort money from groups receiving mining proceeds ("working in the life market"). Other disenfranchised individuals seek wage labor opportunities with variable success or turn to millenarian Christianity in hopes for a better life.

In the concluding chapter I comment on the practical and policy implications of large-scale resource development in Papua New Guinea. Resource exploitation will be a centerpiece of economic development in PNG for the next several decades, and with a combination of foresight and in-depth case studies such as this book hopefully developers and the PNG state will not repeat many of the mistakes that have occurred at Porgera and other mines (Kirsch 2014). I also explore how locally derived ontologies and myths intersect with a resilience framework focused on coupled human and natural systems. I end with a discussion of the agency of gold and how gold mining in particular has lent itself to perceptions of the land ending in Porgera.

Part I

The Making of a Resource Frontier

Porgera is exemplary of the kinds of places that the anthropologist Anna Tsing (2005: 28) calls "resource frontiers," spaces where global capital, multinational extractive corporations, and increasingly militarized Third World countries interact, often on the homelands of tribal and indigenous populations. They are places where local ecologies and livelihoods are disengaged from nature by national governments and where the "natural resources" are offered up as "corporate raw materials." As Tsing notes, these frontiers "aren't just discovered. . . . They are projects in making geographical and temporal experience. Frontiers make wildness . . . [that] is *both* material and imaginative" (28–29). Anthropological concerns with "new frontiers of land control" (Peluso and Lund 2011; see also Fairhead et al. 2012; White et al. 2012) have gained prominence as transnational corporations allied with national elites have increasingly appropriated local people's lands and resources. Peluso and Lund (2011: 668–69) argue that these "new frontiers of land control are being actively created." Building on Tsing's definition, they write, "These created frontiers are . . . sites where authorities, sovereignties, and hegemonies of the recent past have been or are currently being challenged by new enclosures, territorializations, and property regimes. . . . Land control can [also] be understood as embedded within broader political struggles over identities . . . and change[s] the ways we think about enclosure, territorialization, and property as iterative processes productive of environmental subjects." These are important considerations, but what makes the Porgera case so compelling is that it disrupts easy generalizations about how resource frontiers are made. It is not just multinationals, militaries, and national elites that have offered up Porgera's land and resources as raw materials for capitalism, but many Porgerans are also involved in this frontier-making project.

Porgera serves as a microcosm in which to examine contemporary global concerns such as sustainability, deforestation, economic development, resource conflicts, and social-ecological resilience—all the elements involved in making a resource frontier. In chapter 1 I situate Porgera within the larger political economy and ecology of PNG and address my two main themes—the making of a resource frontier and the corresponding implications for social and environmental change—within the theoretical literature on the intersection of society and ecology. From this literature I draw on theories of resilience and change, political ecology, and ontology and place to argue that the mutual constitution of societies and their environments is integral to an engaged environmental anthropology for the twenty-first century. At the end of the chapter I provide a brief ethnographic overview of Porgera as a place and the people who live there.

Porgera is a place that is produced materially, imaginatively, and discursively through the concept of development (cf. West 2006a). Discourses and practices surrounding development stem from two overlapping yet sometimes contradictory forces: resource extraction and Christianity. In chapter 2 I delineate the interactions between colonial agents of development and Porgerans. My goal is to situate the historical process of making a resource frontier from the perspective of Porgerans. How did they understand what they were undergoing, and who benefited and who lost from these engagements? In an attempt to overcome anthropology's focus on the "local" (Wolf 1982), the anthropology of postcolonialism and globalization tends to overemphasize global interconnection. Akhil Gupta and James Ferguson (1992: 8) write, "If one begins with the premise that spaces have always been hierarchically interconnected, . . . then cultural and social change becomes not a matter of cultural contact and articulation but one of rethinking difference through connection." They argue that places are formed out of "interconnected space that always already existed. Colonialism, then, represents the displacement of one form of interconnection by another."

Porgerans live in a multiethnic and multilinguistic social world; they are deeply aware of their interconnections with other groups. However, it is difficult to claim that the larger exchange networks they were involved in during the precolonial era radically reshaped their lives and environments. Today many of them say about the precolonial era that it was a time of "living in a fence"; mobility was restricted and the onset of colonialism either "broke the fence" or "opened the roads," allowing for a greater movement of people. Communities and kin groups were interconnected to an extent in the precolonial era, but not hierarchically.

In chapter 2 I trace the history of "breaking down the fence," from first contact to recent events of large-scale mining. This history is based on documents that I investigated in historical archives, published accounts, and interviews conducted between 1998 and 2006. As I attempt to illustrate, one of the novel effects of colonialism, Christianity, and the postcolonial mining era has been to introduce expanded forms of interconnection that indeed are hierarchical and that profoundly shape Porgerans' sense of their place in the global order. Through interactions with colonial administrators, expatriate miners, and missionaries, Porgerans sought the alchemical transformation of their bodies that I addressed in the introduction, although the outcomes of this process varied greatly from individual to individual. To this end I explicitly document how gold mining interests, implemented in the 1940s, laid the foundation for indigenous mining ventures in the 1950s, which initiated the era of economic inequalities. As a counter to the growing power of the indigenous miners, other men sought alternative avenues to power through engagement with colonial regimes and Christianity. As such, the typical "triad stakeholder" model of the state, the developer, and the community needs more ethnographic attention (Ballard and Banks 2003; see also Agrawal and Gibson 1999; Golub 2014) in order to show how each of these entities is more diverse and factionalized. The aim of the following two chapters is to highlight the diversity and factions at play in Porgera's resource frontier.

1) Land's End Book Review — 1 Book
2) ANTH 4500 Midterm & Essay — 2 Books

3) HIST 4128 — Paper — 3 books

4) HIST 4212 — Paper — IDK A LOT OF
                          READING.

MON — Land's End — 1/2

   4500 — Reading.

4 hrs hw / day
   2 - AM
   2 - PM

[handwritten margin notes:]

situates Porgera within larger political economy & ecology of PNG

address 2 Main Themes:
1) The making of a resource frontier
2) Corresponding implications for social & env. change.

adaptation vs. transformation

CHAPTER I

Resource Frontiers in the Montane Tropics

*[handwritten note:] *shift away from depicting traditional culture & towards dynamic recognition of interrelations between local cultures and global forces.*

Development, Natural Resources, and Conflict

In 2006 Epe Des and I stood on the ridge of a long mountain that bisected the Porgera valley. Dominating the view was the Porgera Gold Mine, one of the world's most productive mines since its development in 1990. Steam from the autoclave where the gold is processed mixed with the low clouds that hid the rain forest–covered peaks surrounding us. Glaciers of waste rock spilled from the mine, snaking past gardens of sweet potato and pandanus trees (figure 1.1). Mining tailings were pouring from the mouth of the mine directly into the Porgera River, which runs blood red as a result of the iron oxide in the mining waste. Each and every day after the ore is crushed and processed another 17,000 metric tons of waste is added to the glaciers of rock. Awed by the immense scale of environmental devastation before us, I silently shook my head. Epe, however, smiled, laughed, and said approvingly, "That's our garden" ("Em i gaden bilong mipela" in Tok Pisin, the lingua franca of Papua New Guinea).

For a subsistence-based horticultural society, a garden is the very life blood that sustains its members. Life for most people in Porgera centers around their sweet potato gardens. While men clear forest, dig ditches, and build fences, women plant, weed, and harvest the crops. There are no seasons in this part of the tropics, nor can one store sweet potato tubers for more than a few days, so gardening is an ever-present feature of daily life. In the morning people wind along the muddy paths to

Figure 1.1. The Porgera Gold Mine.

their gardens and the forest, and in the afternoon women come home with net bags full of sweet potatoes and greens, while men haul long, split lengths of logs over their shoulders for the evening fires. Inside the houses while the sweet potatoes are roasting in the ashes and the greens and other vegetables boil in cooking pots, families talk about the day—what happened in school, whose pig got loose and broke into a garden, where to find money for medicine for a sick child, the current state of affairs in provincial and national politics, and other such topics.

In such a setting what is the significance of Epe's comment? What happens when a world-class gold mine is perceived as a garden—when people "eat" gold? What kinds of new social and environmental relationships are generated by gold extraction and mining development in Porgera? When many Porgerans were engaged in small-scale alluvial mining from the late 1950s to 1990, they called gold their "second garden." The intrusion of state and multinational interests into mining ushered in the demise of gold as a secondary pursuit and has made it the primary road to development in Porgera. When I first arrived in 1998, I expected there to be key segments of society that opposed mining development due to its detrimental impacts on the land and the very subsistence base. I quickly realized, though, that the main problem for Porgerans was that there was not *enough* development. Hopes were high in these initial years of large-scale mining that development would radically transform social life in the valley. In one glossy sustainability report from the mine was an artist's rendering of this Porgeran utopia that showed tree-lined paved streets winding past red tile–roofed Mediterranean-style houses. Adjacent to the town were parks, play fields, and a golf course surrounded by mounded sweet potato gardens. These visions of prosperity have yet to materialize in Porgera. A more accurate vision would capture the multiple images of destruction and burning that are seared into the landscape and history of Porgera due to mining-related violence.

For instance, in 1992 a special police unit assigned to the mine workers' camp killed a fifteen-year-old boy. That night approximately five hundred Porgeran men attacked the camp, burning several vehicles and buildings (Puge 1992). In retaliation, a few months later the special police unit descended en masse onto the town of Paiam, systematically burning "hundreds of houses," stores, and vehicles and shooting pigs (Taylor 1993). In 2006 I documented the burned-out remains of the Porgera Primary School and the charred shell of the former headquarters of the Porgera Development Authority. Both of these buildings had been torched in the preceding months over conflicts related to mining. In 2010 Amnesty International published a scathing report of a massive mobilization of defense force troops and special police units into the Porgera valley under the code name "Operation Ipili," launched by the Papua New Guinean state. The stated goal of Operation Ipili was to quell tribal fighting and the general lawlessness associated with the hordes of illegal miners who sneak into the mine and remove gold from the open pit where PJV has exposed it. Amnesty International reported that from April to July 2009, members of Operation Ipili conducted a harsh campaign of violence around the mining area—burning down houses, beating people, and raping women.[1] Operation Ipili was ostensibly not directed at the Ipili-speaking Porgerans, the indigenous inhabitants of the valley, but rather at the numerous outsiders who had flooded into the region in pursuit of sharing in the proceeds of mining development wealth. In truth, though, determining who is or isn't a Porgeran is not such an easy task (Burton 1991; Golub 2014), and many innocent people suffered needlessly. On March 30, 2012, the front cover of the national paper, the *Post-Courier*, featured the picture of a recently burned Toyota Land Cruiser under the headline "Illegal Miners Attack Porgera Mine." Scorched car parts and shattered glass littered the ground around the destroyed vehicle. While the charred, hulking remains of the car were now windowless, the cages that once covered them to protect the passengers from stone throwers remained intact. The driver's side door hung open, trailing singed weather stripping, providing the image of the driver beating a hasty retreat from an enraged mob of illegal miners. In early 2014 Ben told me that the house we shared in Kolatika had been burned down in tribal fighting; in fact the whole government compound—the symbol of development in Tipinini—had been destroyed.

These images of mining-related conflicts highlight the paradoxes of the alchemy of contemporary resource development. In antiquity, the Golden Age referred to a time of peace, prosperity, and comfort; in the twenty-first century "golden" eras are beset by numerous social and ecological tensions. Central to these tensions is a politics of ecology—struggles over people's relationships and claims to land and resources. Unlike many of the world's indigenous rain forest people, Papua New Guineans maintain customary ownership of approximately 97 percent of the country's land base. In Porgera this means that people in landowning groups where the mine is located not only receive royalty shares and other monetary and indirect benefits from the mine but are partial owners in the stake as well (see Golub 2014). The majority of Porgerans, however, receive few benefits from mining despite having a history of engagement with alluvial mining and long-term kinship and exchange ties to the "landowners." Thus the reason there is not enough development is that the costs of mining are borne by far more people than those who enjoy its benefits. These tensions over enfranchisement and disenfranchisement are not confined to a local context but are also played out at provincial, national, and international scales. Nevertheless it is often local social-ecological systems that bear the brunt of these larger-scale processes (Dove 2011; Ferguson 1999; Tsing 2005).

The Ecological Context

The third largest extant block of rain forest in the world exists on the island of New Guinea, yet it receives far less media attention as an endangered forest than the rain forests of the Amazon or Congo. The forests of New Guinea, Indonesia, the Philippines, and peninsular Malaysia, a region ecologists call Malesia, harbor approximately forty-five thousand plant species, with the richest flora likely to be in New Guinea, although the island has received very little scientific focus (Primack and Corlett 2005: 34). Approximately 40 percent of the plant genera are endemic to New Guinea and Southeast Asia (Whitmore 1984: 5), and the rain forests are much richer in tree species than either the Amazon or the Congo (6). New Guinea is the world's highest-altitude island (several peaks over 4,800 meters), the second largest island in the world and the largest in the tropics, and one of only three tropical areas with glaciers. The ecologist J. L. Gressit (1982: 3) described New Guinea as "a fantastic island, unique and fascinating. It is an area of incredible variety of geomorphology, biota, peoples, languages, history, traditions and cultures." As a consequence of its biological uniqueness, in 2003 Conservation International designated the island one of the planet's five High Biodiversity Wilderness Areas given its large expanse of intact rain forest, low human population density, and high numbers of endemic species of flora and fauna (Mittermeier et al. 2003). The island is just as culturally and linguistically rich, containing approximately one-sixth of the world's linguistic diversity.

The eastern half of the island of New Guinea and several surrounding archipelagos constitute the nation-state of Papua New Guinea. With a large land mass (around 460,000 km²) and a relatively small population (approximately 7 million), the tropical rain forests of PNG are still extensive, yet within the past decade they have been subject to one of the highest rates of clearance in the world (FAO 2010). Today over 60 percent of the land area is covered in mature rain forest, down from approximately 70 percent in the early 1970s (Shearman et al. 2008). With the contemporary worldwide emphasis on reducing tropical deforestation and minimizing carbon emissions to combat global climate change, the PNG state is under mounting international pressure to preserve and protect its forests. As a consequence, since the mid-2000s PNG has occupied a major international role in promoting the preservation of its rain forests through participation in the United Nations Collaborative Programme on Reducing Emissions from Deforestation and Forest Degradation in Developing Countries and other international carbon initiatives (Filer and Wood 2012), all of which seek to reduce emissions from deforestation and forest degradation in developing countries.[2]

At the same time that global conservation initiatives and segments of the PNG state have sought to protect the rain forests, the island of New Guinea is also experiencing a mining, logging, and petroleum development boom that promises to acutely impact the quality and extent of forests. The island is home to several world-class gold, copper, and nickel mines; many of the lowland forests are being logged under timber concessions; and numerous petroleum deposits are under development on the eastern half of the island. Currently a 9-trillion-cubic-foot liquid natural gas deposit is being developed in PNG that is expected to boost the country's GDP by 40 percent and its export revenue by 67 percent per annum (Dixon et al. 2010: 143). Overall the mining and petroleum sectors contribute to approximately 70 percent of Papua New Guinea's export earnings, 30 percent of its total national revenue, and 25 percent of its

GDP (Togolo 2006). Despite the wealth that these development projects bring to the state, most citizens see little in the way of these benefits. In 2014 PNG was ranked 157 out of 187 countries by the United Nations Development Programme's Human Development Index that assesses three measures: health, education, and living standards. For purposes of comparison, this puts PNG in the same category of development as most of the countries in sub-Saharan Africa. Approximately 85 percent of the population is dependent on the forests and the natural environment for their livelihoods through subsistence and semisubsistence agriculture (Bourke and Harwood 2009). Yet these same rural people clamor for improved livelihoods from development projects that are often the very source of large-scale environmental degradation, like logging, mining, and plantation agriculture.

Two earlier mining projects in PNG foreshadow some of the issues surrounding the Porgera Mine. The first mine in PNG, the Panguna copper mine, which opened in 1972 on the island of Bougainville, precipitated a civil war in the late 1980s as the Bougainville Revolutionary Army undertook militant actions to close the mine and secede from PNG (May and Spriggs 1990). Locals were upset over the environmental devastation, riverine pollution, and royalty payments and succeeded in stopping mining operations in 1989. Panguna, once the largest copper mine in the world, provided 44 percent of the export revenue for the PNG state during its seventeen years of operation, and its closure shocked the PNG mining industry (Bougainville Copper Limited 2013). With 3.5 million metric tons of copper and over 12 million ounces of gold still in the ground,[3] the PNG state and mining developers are eager to get the project going again but face opposition from some landowners in the area. PNG's second major mine, Ok Tedi, has faced nearly as many problems as the Panguna mine, although it continues to operate. Opened in 1984, Ok Tedi, like Panguna, was permitted to dump tailings into local rivers. The increased sediments and toxic metal loads in the Ok Tedi and Fly rivers have been devastating. In 1984 a massive cyanide leak killed riverine flora and fauna for over 100 kilometers downstream from the spill (Kirsch 2006: 16–17). Due to the increased sediment loads, people living downstream from the mine have suffered siltation of the river beds, detrimentally impacting subsistence and travel on the rivers, and have had gardens destroyed due to the rivers spilling over their banks from sedimentation (Kirsch 2006). In 1994 the mine was sued on behalf of thirty thousand indigenous landowners for environmental damages, and two years later, a K110 million out-of-court settlement was awarded to local landowners (see Banks and Ballard 1997; Kirsch 2014).

Consequently the challenges for rain forest nations like Papua New Guinea that try to balance economic development with ecological conservation are immense (West 2006a). While some sectors of the state and the populace are working toward developing resources for socioeconomic improvement, others are seeking the preservation of resources and hoping that ecosystems services, such as carbon credit trading (REDD+), ecotourism ventures, and integrated conservation and development projects, will lead to improved social, economic, and ecological outcomes. At the center of these global discourses on biodiversity conservation and economic development, although often excised or hidden from view (Dove 2011; Tsing 2005), are local people whose livelihoods and cultures engage in substantive ways with these resources. A central concern for a critical, engaged environmental anthropology is to bring to light these subjugated worldviews—"to take into account indigenous understandings of knowledge and practice" regarding how people interact with plants, animals, their landscapes, and one another (West 2005: 633; see also Cruikshank 2012; Descola 2013; Nadasdy 2007; Viveiros de Castro 1998, 2004). To complement this, we also need to position these "indigenous understandings" in line with the global expansion of capitalist economic forces to better understand how diverse cultures and their environments shape and are shaped by global processes.

Resilience and Change

In the early 1980s anthropology took a historical (Sahlins 1981) and political economic (Mintz 1985; Wolf 1982) turn as it attempted to account for the ways that local people were entangled in the larger global system. The shift was away from static depictions of traditional culture to a more dynamic recognition of complex interrelations between local cultures and global forces (Appadurai 1996; Gupta and Ferguson 1997). In some ways these shifts in cultural theorization were in parallel with what was happening in studies of ecology. In the late 1960s researchers like Eugene Odum (1969) described ecosystems as being self-regulated by homeostatic processes that ensured that they remained in stable states or had mechanisms that ensured they quickly returned to those states. Change was linear and followed Clements's (1916) theory of succession to climax ecosystems. Others, like C. S. Holling (1973: 15), argued that equilibrium states were rarely found in real-world situations and that instability suggests the importance of "resilience and a capacity to persist" for species in complex systems. Dynamic change was central to Holling's notion of resilience, as opposed to equilibrium concepts of resilience, where change is subverted by regulatory mechanisms that preserve homeostasis (Rappaport 1967). I explore these ideas thoroughly in part III.

Recent developments in the social and ecological sciences urge us to view change as a normal process and not an aberration. From social science Marshall Sahlins provides some seminal ideas for thinking about how to study change. Sahlins (1999) urges anthropologists to study "the indigenization of modernity" to recognize that so-called cultural traditions are dynamic and always changing. People are not refusing "the commodities and relations of the world system" but "desire to indigenize them" and make them their own (x; see also Sahlins 1994; Wolf 1982). In this view economic globalization rediversifies the world through "indigenous adaptations to the global juggernaut" (Sahlins 1999: ix) and allows people to develop "their own culture on a bigger and better scale than they ever had it" (Sahlins 1992: 13). Cultural change is a given and not to be lamented and misconstrued as cultural loss (what Sahlins 1999: iii calls "despondency theory").

Likewise, from a human ecology perspective, Ian Scoones (1999) calls for social scientists to engage more dynamically with findings from the new ecology, with its focus on spatial and temporal variation, nonequilibrium dynamics, and complexity (see also Gunderson and Holling 2002). Scoones (1999: 490) highlights three themes around which such work should coalesce: (1) "people in places," which uses historical analysis to explain environmental change across space and time; (2) the "environment as both the product of and the setting for human interactions"; and (3) "the appreciation of complexity and uncertainty in social-ecological systems." Both Sahlins and Scoones draw our attention to the critical notion that change is a consistent element of social-ecological systems, a notion that is also central to a resilience framework (Berkes et al. 2003a).

This highlights that we need to define more clearly what kind of changes we are trying to analyze. Is change a matter of degree (Sahlins 1981), in which small shifts in social-ecological systems are readily absorbed? Or is change a matter of kind, in which there are large-scale transformations in social-ecological systems that require significant responses to accommodate them or that end up radically changing the system? Peter Dwyer and Monica Minnegal (2010: 632) attempt to define these two types of change, respectively, as "adaptation" and "transformation." Adaptation is when shifts occur without "substantive alteration to functional relationships" in a social-ecological system. Transformation occurs when there are substantive alterations that impact "the structure of the ensemble as a whole." They argue that only in the latter type of change will "new adaptive responses . . . arise." Interestingly, while ecologists and sociocultural anthropologists were examining change, the majority of environmental social scientists were still very much embedded in concepts of stable, unchanging ecosystems and cultures in adaptation with them (Scoones 1999).

Applying resilience to social-ecological systems, Holling and colleagues (2002a: 21) argue that these systems are constituted by linked economic, ecological, and institutional processes (see also Berkes et al. 2003b; Berkes and Folke 2000b; Gunderson and Holling 2002; Norberg and Cumming 2008). The concept of resilience offers much to an environmental anthropology that embraces change through the fact that resilience frameworks highlight "two opposites: growth and stability on the one hand, change and variety on the other" (Holling 2001: 395). Additionally resilience offers a way to incorporate ecological analyses in political ecology (Vayda and Walters 1999) and provides a way for environmental anthropology to speak across interdisciplinary divides (Dove 2006). However, I argue that rethinking and expanding notions of economics and institutions can strengthen resilience approaches. As currently conceived in the resilience literature, all of the various aspects of "culture" that anthropologists study are reduced to "institutions," while all of the dynamic elements of political economy are reduced to individualistic and rationalist cost-benefit analyses. In place of neoclassical economics in resilience studies (Scheffer et al. 2002), political economic approaches that are central to political ecology would provide a much needed foothold in processes of globalization that are missing in a resilience framework. Institutions (Agrawal 2003; Ostrom 1990) are merely one part of resource management systems within different cultures. Other factors, such as belief systems, indigenous ecological knowledge, and worldview, have been largely disregarded by most of the research operating in a resilience framework (but see Berkes 2008; Berkes et al. 2000, 2003a; Berkes and Folke 2000a).

Carpenter and colleagues (2001) stress that when studying resilience it is crucial to emphasize "resilience *of* what" and "resilience *to* what." Much of the way I define resilience in the context of this book stems from the perspectives of Porgerans themselves in terms of impacts on their ontology and livelihoods (e.g., subsistence practices, social networks, quality of life). I document where they perceive radical transformations in their social-ecological systems and where they perceive continuity and stability. The focus is on the resilience *of* social-ecological systems as understood by Porgerans, and resilience *to* various environmental and social crises as defined by Porgerans, such as changes in soil fertility, climatic fluctuations, human well-being, and so forth. To clarify, Porgeran livelihood practices center around the production of sweet potatoes through shifting cultivation, a process that mimics the formation of gaps in tropical rain forests by natural means (Bayliss-Smith et al. 2003; Whitmore 1989). At the temporal scale of one season and the spatial scale of the garden, Porgeran subsistence is resilient in the face of nutrient-deficient tropical soils and the challenges of a tropical climate (Sillitoe 1996). However, given the montane environment of Porgera, landslides, temperature and rainfall variability, and differences of soil quality, there is a need to maintain multiple gardens to ensure sufficient food over a longer time span. At larger spatial and temporal scales, resilience is still found in the Porgeran food production system. At times, though, individuals, lineages, and even entire clans perceive that there are declines in soil fertility and crop production (i.e., "crises") that necessitate actions beyond the mundane and routine, such as ritual propitiation to ancestral and other spirits. Today especially mining development generates even larger crises in Porgeran livelihood practices and their locally derived ontology. These crises represent perceived potential collapses in the social-ecological systems of Porgerans. Examining what Porgerans portray as a crisis and how they respond to such an event allows us to understand resilience as internally defined rather than as externally imposed by some measure such as biodiversity, percentage of forest cover, or some socioeconomic metric.

Political Ecology

In the middle of the twentieth century, Julian Steward (1955) founded the first distinctly environmental subfield of anthropology, which he called "cultural ecology." The central focus of cultural ecology was understanding how particular cultures adapted to their environments by studying cultural "features which are most closely related to subsistence activities and economic arrangements" (37). Steward's contention with mainstream anthropology at that time was that it stressed "that historical factors are more important than environmental factors, which may be permissive or prohibitive of culture change but are never causative. . . . The thesis of this paper [is] that cultural ecological adaptations constitute creative processes" (34). In making this statement about creativity, Steward was allowing for a perspective of adaptation as nonmechanistic and was highlighting the potential for exploring both the material and symbolic dimensions of adaptation. While cultural ecology was eventually criticized for its units of analysis (Rappaport 1967), its emphasis on cultural evolution (Orlove 1980), and its failure to recognize the dialectical relationship between society and environment (Baleé 2006; Crumley 1994), the materialist emphasis on studying human uses of the environment in production and subsistence activities is still central to most current trends in environmental anthropology (for a comprehensive history, see Paulson et al. 2005). Subsequent research, though, emphasized the need to integrate other cultural factors—besides subsistence—into human ecological practices. The critical text in this regard was Roy Rappaport's (1984) book *Pigs for the Ancestors*, first published in 1968. Rappaport's analysis centered on the Tsembaga Maring of highlands Papua New Guinea and argued that ritual was the principal mechanism that regulated human-ecological affairs. A massive pig killing ritual, called *kaiko*, nourished ancestral spirits and allies who assisted the Tsembaga during warfare. While the Tsembaga saw the basis of the rituals as realigning the relations between the people and their ancestral spirits (Rappaport 1984: 6–7), Rappaport argued that the kaiko had significant ecological effects in terms of regulating population sizes of humans and pigs, conserving wild game species, mitigating warfare, and redistributing agricultural land and other goods (3–4). The significance of Rappaport's work lies in its interdisciplinary borrowing, especially from systems theory and nutrition science (Dove 2006). At the same time, though, this borrowing opened up his work to serious critique and was described by his detractors as "vulgar materialism" (Friedman 1974) and "ecology fetishism" (Sahlins 1976), among other things. His work was subject to more criticism as anthropology turned away from positivism toward "interpretation" (Geertz 1973) and history and political economy (Wolf 1982). However, in a subsequent reanalysis of the importance of *Pigs for the Ancestors* for contemporary environmental anthropology, Aletta Biersack (2006b: 7) argues that in retrospect the book is an "ethnography of nature . . . a study of the intersection of culture and nature, rooted as this intersection is in human activity, conceptualization, values, and social relations."

The turn toward political economy in environmental anthropology stresses important concerns about the role of capitalism in the intersection of social and biological worlds. The environmental sociologist John Bellamy Foster (2000: 155–70) has argued that we need to attend to the ways that capitalism creates "metabolic rifts" between social and natural worlds, or in other words, works to delink nature and culture. As Marx (1973: 489) wrote in *Grundrisse*, "It is not the unity of living and active humanity with the natural, inorganic conditions of their metabolic exchange with nature, and hence their appropriation of nature, which requires explanation or is the result of a historic process, but rather the *separation* between these inorganic conditions of human existence and this active existence, a separation which is completely posited only in the relation of wage labour and capital." Marx (1990: 283, 290) also wrote about metabolism in volume 1 of *Capital*, where he describes labor as the process through which humans mediate, regulate, and control the metabolism between "[themselves] and nature . . . as a force of nature" and as "the everlasting nature-imposed condition of human existence." Capitalism, especially industrial agriculture, creates a metabolic rift through the "robbing of the soil of the countryside of nutrients and the sending of these nutrients to the cities in the form of food and fiber. . . . This rupture in the soil nutrient cycle undermined the regenerative capacity of the ecosystem" (Foster et al. 2010: 45–46). To better understand how these processes operate in the resource frontiers of the developing world, it is essential to explore the ways that social and ecological rifts form in the interstices of capital and ecology.

While political ecology emphasizes the importance of understanding local and extralocal interactions as key mechanisms in producing contemporary human-ecological relationships, it is also important to realize that not all human-ecological interactions are generated from local and extralocal tensions (Vayda and Walters 1999; West 2005). Steward's claim about cultural creativity as a form of adaptation, Rappaport's ideas of the intersection of ritual and ecological systems, and Biersack's concept of an ethnography of nature take as their central assertion the understanding that society and the natural world compose a mutually constitutive relationship (see also Biersack 1999a; Descola and Pálsson 1996). As Tim Ingold (2000: 19) has written, this mutual constitution needs to be seen as an "emergent" form that is produced as a direct result of our engagement with the world. This is especially relevant for our knowledge of how places are "products of and settings for" (Scoones 1999: 490) human-ecological interactions (Bebbington and Batterbury 2001).

Ontology and Place

Places, according to Keith Basso (1996b: 53), are "cultural constructions of geographical realities." Cultural concepts of place are exemplars of this creative space at the interface of cultural and natural systems where human systems of meaning engage with biophysical realities (Escobar 1999). Yet, as Margaret Rodman (1992: 641) reminds us, "places are not inert containers. They are politicized, culturally relative, historically specific, local and multiple constructions." Studies of place-making allow us to work through culture-nature dichotomies by emphasizing the mutual constitution of societies and natures (Descola and Pálsson 1996). The concept of place also provides a way for us to explore the articulations between local and global processes central to place-making (Biersack 2006b; Dirlik 2001), for places are not just locally constructed but are the result of multiscalar interactions affecting social-ecological systems. Critical perspectives on place (Biersack 1999a, 2006b; Escobar 2001, 2008; Gupta and Ferguson 1997; Massey 1994) thus provide an important framework for encompassing theories of globalization through what K. Sivaramakrishnan and Arun Agrawal (2003: 21) call "the regional production of the experience of globalization," which describes the "networks, struggles, and differentiated place-making that people recreate in everyday contexts." Studies of how places are socially produced and culturally constituted highlight the "micro-politics" (Moore 1993) of social experience as men and women, young and old, and elites and commoners struggle to create meaningful lifeworlds. Describing how place is produced, maintained, fought over, and experienced allows us to weave together issues of spatial and temporal scale addressed in the resilience literature. In fact the philosopher Edward Casey (1996: 36) argues that "the phenomenological fact of the matter is that space and time come together in place." They come together because "place is the most fundamental form of embodied experience" (Feld and Basso 1996: 9).

As embodied experience, place highlights a critical fusion between human productive activities and the perception of those products (West and Carrier 2010). Place allows us to understand how materiality and perception are crucial elements in studies of social-ecological systems. Materiality has been described as "the ontological reality of these entities we term 'natural,' and the active role those entities play in making history and geography" (Castree 1995: 13). In other words, materiality refers to a nature that is produced yet still has its own agency and causality (e.g., Bakker 2004). It is closely related to Pierre Bourdieu's (1977: 78) concept of habitus—"history turned into nature"—which combines Marxist materialist analysis with phenomenological concepts of embodiment, dwelling, and movement (Ingold 2000). Perception is not just knowledge; it is "synesthetic—an affair of the whole body sensing and moving" (Casey 1996: 18). Perception therefore is active knowledge; it is produced by activity within a particular landscape (Viveiros de Castro 1998). In discussing the relationship between place and meaning, Tim Ingold (2000: 192) writes, "A place owes its character to the experiences it affords to those who spend time there—to the sights, sounds and indeed smells that constitute its specific ambience. And these, in turn, depend on the kinds of activities in which its inhabitants engage. It is from this relational context of people's engagement with the world, in the business of dwelling, that each place draws its unique significance. Thus whereas with space, meanings are *attached* to the world, with the landscape they are *gathered from* it." Meaning about place is therefore derived from practical engagement with the world. Understanding how different cultures understand "that world" leads us to ontology.

The recent ontological focus in anthropology urges us to consider the kinds of collectives that various societies create from the stock of humans and nonhumans at hand. In an attempt to analytically structure these collectives, Philippe Descola (2013) argues that there are two basic properties that define how people perceive others: interiority and physicality. Interiority is mind, consciousness, emotions, and other such qualities, while physicality refers to bodies and outward appearances. All societies can be categorized by whether they assign attributes of sameness and difference to interiority and physicality, which allows for four ontologies. The first, naturalism (Western ontology), posits physicality as the same for humans and nonhumans, while interiorities are different. In other words, humans and other biological and abiotic entities are all constrained by the same laws of nature. We (Western-educated people) share an animal nature with the nonhumans in our environment, while we reserve humanity (interiority) for humans. As Descola notes, however, Western science's foray into studies of chimpanzees, dolphins and whales, and other animals is starting to disrupt the notions that humans alone have culture, learning, and consciousness (172–200). The second ontology, animism, is the inverse of naturalism. From this perspective, interiority is the same between humans and nonhumans, while physicality is different (multinatural). Animistic ontologies attribute personhood and subjectivity to nonhumans (see Reichel-Dolmatoff 1996; Willerslev 2007), claiming that what differentiates humans from nonhumans is our bodies. Totemism, the third ontology, assumes that the interiorities and physicalities of humans and nonhumans are similar, while the fourth, analogism, views their interiorities and physicalities as different. In Descola's schema, classic totemic societies are those of Aboriginal Australians represented by Dreaming, while ancestor worship, sociocosmic hierarchies, and sacrifice characterize societies with analogic ontologies.

While Descola's structural typology of ontologies provides a critical means to analyze the kinds of relationships humans have with nonhuman entities in their environments, its application to Porgeran ontology is problematic. By nearly all measures, as will be seen in the following chapters, Porgeran culture would best be encapsulated by analogism, as they sacrificed to ancestors and autochthonous spirits ordered by a spatiotemporal hierarchy (see chapters 3 and 4). At the same time, though, humans share a similar substance—*ipane*, or grease—with other living beings and even the land. Thus their interiority is similar to the biotic and abiotic entities in their landscape. Subjectivity is granted to nearly all entities that are motile and hence make willful decisions about where to go (see chapter 5). These latter attributes would therefore place them firmly within an animistic ontology. Despite this problem of fit, Descola offers an important contribution to the study of ontology by focusing on the question of the ways that interiority and physicality are similar and dissimilar across various societies. I agree with Eduardo Viveiros de Castro's (2004: 484) declaration that "we need richer ontologies" to extricate ourselves from the naturalist fallacy that we are separate from an external environment that we think of as "nature." Not only do we need richer ontologies, but we also need to attend to the impacts of ontological change. How do people experience and understand disruptions (ecological, social, ideological) in their lived-in worlds? What are the lingering effects of ontological dissonance when indigenous ontologies are juxtaposed to naturalist and Christian ontologies? With these theoretical concepts in mind, I turn to the place and people of Porgera.

Porgera: Place and People

The Porgera valley is located in western Enga Province (5°16' to 5°33' S by 143°0' to 143°16' E) where heavily forested, steep mountains cut by tumbling, rushing rivers form a series of interconnected V-shaped drainages approximately 553 square kilometers in area (Map 1.1). In the southern part of the valley, limestone mountains over 3,800 meters above sea level form an almost impenetrable wall nearly a thousand meters high. At the base of this wall, houses and gardens are scattered throughout the forest between about 1,600 and 2,400 meters. In the lower reaches of the northern Porgera valley from 1,600 to 900 meters where it flows into the Lagaip River, malaria-carrying *Anopheles* mosquitoes preclude human occupation. Porgera's geology comprises deep beds of limestone layered between sedimentary mudstones and siltstones, a result of the New Guinea landmass forming from sediments eroded from the ancient southern continent of Gondwana in addition to long periods when much of New Guinea lay under seawater (Gurnis et al. 1998). New Guinea's location along the Pacific Ring of Fire results in a complex geological setting

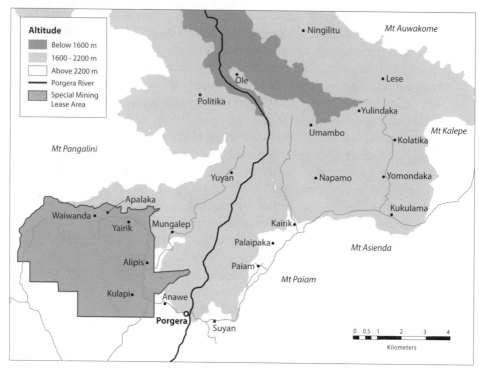

Map 1.1. Hamlets in the Porgera valley.

with frequent earthquakes, faulting, and volcanism (Gunson et al. 1997; Löffler 1977). Around 6 million years ago, gold-enriched magma flowed into a series of faults to within 2 to 3 kilometers of the earth's surface (Gunson et al. 2000; Richards et al. 1997; Ronacher et al. 2000; Upton and Gow 1998). The Porgera Intrusive Complex, as geologists now call the ore body, ceased intruding between 1 million and 300,000 years ago. Subsequent faulting, folding, and erosion moved the gold to the surface, allowing for its discovery in 1939, precipitating many of the events that unfold in this book.

Volcanic activity to the east of Porgera in the geological past has spread eruptive products over much of the area (Löffler 1977: 24), resulting in relatively young, fertile soils in the region (Sillitoe 1996). These volcanic ash soils are the most important soils in local agricultural systems, as they are more resistant to erosion than soils derived from other sources (Wood 1983: 72). Erosion and landslides are a challenge to subsistence agriculture. High annual rainfall inundates the siltstones and mudstones and penetrates the porous limestone geology.

As a consequence numerous waterfalls gush from holes midway down the steep cliffs that form the valley's mountainous southern border. Elsewhere in the high-altitude limestone karst country above these mountains, entire rivers disappear into sinkholes only to emerge kilometers away. The geology creates a landscape pockmarked with caves, sinkholes, and underground rivers (in the limestone), with soils highly susceptible to erosion and landslides (in the siltstones and mudstones). Consequently Porgerans have a rich vision of their underworld as an intricate network of caves and tunnels through which spiritual entities move and in which they live. The frequent landslides and earthquakes are strong evidence for the existence of these entities. Due to the various inundations of the island of New Guinea during its geological history, there are numerous marine fossils embedded in the limestone that is used as gravel for surfacing the roads and markets in the valley. People are intrigued by the presence of fossils and see in their existence the validation of flood stories, such as the biblical flood of Noah, and the adventures of a mythical ancestor-hero, Lemeane, who created the eastern Porgera valley with the help of his giant pig (see chapter 5).

The rain in Porgera is so predictable that every afternoon around 3 or 4 p.m. the *alua* bird (White-throated Honeyeater, *Melithreptus albogularis*) calls out, "Kingi epeya, kingi epeya" (The rains are coming). On one particular afternoon, working with Epe and his family in their garden, Epe's wife, Isam, tells us all to quit working. "Kingi epeya," she says. "Let's leave before we get soaked." The women hoist large string bags filled with sweet potatoes over their back and position the straps across their forehead. Digging spades in hand, Isam and her daughters, Angie and Rose, and Sandam, a niece, set off down the trail. Epe, his son David, and I grab the machetes and follow the women to the trail back home. On this afternoon, as on most afternoons, the alua is right. Before we have gone even a few hundred meters, lightning flashes and thunder peals off the sheer cliffs of Mt. Asienda above us. We hear the intensifying roar of the rain on the leaves of the trees several seconds before it hits us. And then the rain comes; in no time our clothes are soaked. Water drips off the tip of my nose with each step I take. Soon the trail is a muddy rill, and my gumboots slip and slide along the clayey track. I envy Epe and his family's ability to go barefoot, as their feet grip the roots and rocks that I slip and slide on. For most Porgerans shoes are too precious a commodity to be worn to the garden and are saved for trips into town.

Whereas the intensity of the rainfall in the New Guinea highlands is less than many other tropical regions, its abundance surpasses most places on the planet (Primack and Corlett 2005: 10). Rainfall data from

Porgera between 1974 and 2011 show that on average the region receives 3,741 millimeters per year.[4] While the rain forests of the Amazon and Congo basins annually experience dry seasons (periods of several months in which rainfall does not exceed 100 millimeters per month), in only six months of thirty-eight years of data did rainfall not exceed 100 millimeters in Porgera. Four of these months occurred in the middle of two of the most significant El Niños in the twentieth century (1982 and 1997), while the other two months occurred at the end of the 2010 El Niño. Between 1974 and 2011 the minimum annual rainfall was 2,506 millimeters (the 1997 El Niño year) and the maximum was 4,596 (in 2011). One group of climatologists writing about seasonality in Papua New Guinea referred to the seasons as "wet and very wet" (McAlpine et al. 1983: 61), as opposed to monsoonal wet and dry patterns that are found farther from the equatorial zone. There is a peak in the "very wet" period extending from October to April and a less wet period from May to September. The Ipili terms *kingi* (rain) and *panyu* (good weather) are used to refer to seasons, but these terms only partially pertain to the recognition of seasonality, as any time throughout the year that it is not raining, people say that they are experiencing a panyu, a dry season, which often lasts only a few days. Moreover cropping patterns do not indicate seasonality, as new gardens are planted throughout the year.

Based on analysis of Landsat data from 2013, approximately 70 percent of the southern Porgera watershed is covered in primary rain forest (see chapter 6 for more details).[5] In PNG botanists (Johns 1982) distinguish between lower montane "Castanopsis" forest and midmontane "Nothofagus" forest. Lower montane rain forest is found below about 2,000 meters and is dominated by oaks (*Castanopsis acuminatissima* and *Lithocarpus* spp.). Lower montane rain forest has a mean canopy height of 30 to 40 meters, with 70- to 80-meter emergents of hoop pine (*Araucaria cunninghamii*) in scattered stands below about 1,800 meters. Midmontane rain forest, dominated by southern beech (*Nothofagus* sp.) and *Pandanus* spp., occurs from about 2,000 to 3,000 meters and has a mean canopy height of 20 to 30 meters, with 30-meter emergents of Nothofagus. Upper montane (also called cloud or mossy) forest and naturally occurring grasslands and tree fern (*Cyathea*) savannas occur above 3,000 meters. Approximately 25 percent of the southern watershed is composed of secondary forest and gardens. At mid- and lower elevations, oaks dominate in secondary forests, while regrowth forests in upper elevations (2,200 to 2,600 meters) are dominated by tree ferns and *Homolanthus nervosus*. The mine and built-up areas make up the final 5 percent of the land area.

Porgerans, like other highlands New Guinea peoples, practice shifting cultivation. However, due to soil-enhancement practices, such as mounding and composting, gardens can be productive for more years than shifting cultivation that is done without these practices (Sillitoe 1996). The primary staple is sweet potato (*Ipomoea batatas*) that is grown in composted mounds. Mounds are multicropped with maize (*Zea mays*), pumpkins (*Cucurbita* spp.), onions, potatoes (*Solanum tuberosum*), tomatoes, beans, greens (*Amaranthus* spp., *Rorippa* spp., choko [*Sechium edulis*] leaves and pumpkin leaves), highland pitpit (*Setaria palmifolia*), and sugarcane. Bananas (*Musa* spp.) and taro (*Colocasia esculenta*) are important secondary staple crops. Various pandanus tree (*Pandanus julienetti*) varieties are planted in gardens for their nuts, which provide an important source of protein and fat (Hyndman 1984; Peters et al. 1960; Stone 1982). Domestic pigs (*Sus scrofa*), however, provide the most common source of fat and protein. Goats and chickens are becoming increasingly important domestic animals as well.

Wild fauna provide meat for consumption and skins or feathers for personal adornment. There are a wide array of game animals, various species of tree kangaroos, quolls, cuscus, possums, and bush rats, as well as echidna (*Zaglossus bruijni*) and wild pigs (*Sus scrofa*), that are hunted. Possum, cuscus, and tree kangaroo skins are used for headdresses worn during ceremonial exchange events. Dwarf cassowaries (*Casuarius bennetti*) are the most important of the bird species that are hunted for game. Their tail feathers are used in headdresses and their claws are used as spear tips. Birds of paradise, cockatoos (*Cacatua galerita*), lories, lorikeets, and various pigeons and doves are hunted for their feathers, which are used for personal adornment and exchange. The nests of megapodes (Megapodiidae) are raided for the abundant large eggs they produce. With the influx of population from mining, many of the wild fauna have been extirpated from areas near the main road, although hunting remains a culturally important practice. Various fungi, wild fruits, and medicinal plants are gathered from the forest.

Most people enter the Porgera valley via the Enga Highway, a two-lane gravel road that stretches from Wabag, the provincial capital, to Porgera. The first settlements along the road into Porgera compose a region called Tipinini, made up of several hamlets in the eastern Porgera valley. Past Tipinini is the hamlet of Kairik, where the new airstrip was built in the late 1980s (figure 1.2). A few kilometers past Kairik the highway passes the town of Paiam, built as part of an agreement for mining development to proceed (Jacka 2001, 2007). Paiam has a public market, the local high school, a hospital, a rugby field, some government buildings, a large

Figure 1.2. Dancers on the Enga Highway en route to a *singsing* next to the Kairik airstrip.

grocery store, and a Toyota dealership. Some of the mine workers and government employees live in houses behind razor wire–topped fences. Beyond Paiam is Suyan, the main camp for housing the mine workers. After the riot in 1992 that damaged many of the mining camp facilities, PJV surrounded the camp with a second ring of razor wire–topped fences and guard towers along the inner perimeter. Just past Suyan is the center of Porgeran social life, Porgera Station—the Station, as most people call it, Yanjiakale to older people who still use the precolonial name, or just "Porgera" some of the time. In 1961 the Australian colonial government built a crushed-limestone airstrip and established a permanent patrol station to administer the local area. Today the airstrip is no longer in use, and the public motor vehicle (the main means of transportation in PNG) stop now cuts it in half. Past the Station the road continues on to the mine and the landowner communities surrounding it.

Lining the edges of the old airstrip at the Station are corrugated-iron stores with barred windows, many of them enclosed in compounds with chain-link fences topped by razor wire. Security guards stand at the entrances; to deter theft they pat people down and check their bags when they leave. Every third or fourth store has a sign outside it, *Baim Gol Hia*

(Gold bought here), where alluvial miners (most panning in the mine tailings) can sell their gold and purchase panning supplies and vials of mercury, used to separate the gold from the sand. Outdoor markets anchor the upper and lower ends of the airstrip. The upper market, adjacent to the Porgera District Administration headquarters, sells mostly produce, skeins of colorful yarn that women use to weave net bags, and cheap trade goods from China. The lower market, the *buai maket* (betel nut market), is the center of illicit and semilegal activities in Porgera. During my initial fieldwork period it was the source for illegal liquor. Rows of dartboards used for gambling line the buai market—"Hit the bulls eye and win a Coke"—while groups of people squat in the red betel nut–stained mud, gambling with cards. Pockets of loud, raucous women, most of whom are reputed to be *pasindia meri* (passenger women; women who accept money for sex; see Wardlow 2004), laugh and yell and flirt with the mostly male visitors wandering around the market.

"Wandering around," *raun raun* in Tok Pisin or *me ata* in Ipili (which also means "doing nothing"), is perhaps the best way to explain what people do at the Station (see chapter 7). At any given time there are hundreds of people just standing, sitting, or strolling along the airstrip, its surrounding businesses, and the government offices. In the stores people buy a few small items, some soap, a single cigarette, maybe some cooking oil, nothing major. Since the late 1980s, when mining development started, the bulk of the population in the valley moved from the middle reaches of the valley to higher altitudes to be closer to the highway and the Station. The land is poorer along the road and gardens are less productive for sweet potato production. As most of the road is above 2,200 meters above sea level, taro—an important secondary staple—is unable to grow in nearby home gardens. As a consequence many people maintain small-scale business ventures, selling betel nut and single cigarettes or pieces of fried wheat dough (*flaua* in Tok Pisin) or mutton (*sipsip*). The highway communities are also more dependent on store-bought goods like rice, packet noodles, and tinned fish to supplement their less productive gardens. As local mine workers are not housed in the mining camp, wage incomes in these communities help to offset the added cost of purchasing food.

As mentioned, the anthropological literature (some of my own included) has called the people of Porgera "the Ipili."[6] In this book I refer to the contemporary inhabitants of the valley as Porgerans, as *the Ipili* elides several important facts about the social, ethnic, and linguistic composition of the valley's inhabitants today. The contemporary situation in Porgera is a microcosm of the country's linguistic diversity of over eight hundred languages, with "Porgerans" speaking a number of different languages depending on whether they were born in the valley or have migrated in from elsewhere to be near the development associated with mining. Prior to mining, most of Porgera's population were Ipili speakers, and it is estimated that there were between 4,000 and 5,000 Ipili living in Porgera during the 1970s and 1980s. By 1990, the first year the Porgera Gold Mine was in operation, there were 9,255 people living in the valley. In the 2000 census, there were 22,809 people in Porgera, and mining officials estimate that in early 2008 there were about 42,000 people in the valley (Gimpel 2008). Assuming that the annual growth rate in population of 9.4 percent between 1990 and 2000 is still valid, or the 7.0 percent rate from 2000 to 2008 is, there could be between 50,000 and 70,000 people living there today.[7]

Language Groups around Porgera

According to Ethnologue (Lewis et al. 2013), the online database of the world's languages, there are 26,000 speakers of Ipili. To the east of the Ipili are the Enga, around 230,000 speakers, and to the south are the Huli, with around 150,000 speakers. To the north of Porgera are two other language groups, Hewa and Nete, each with around 1,000 to 2,000 speakers. To the west of the Paiela valley Ipili are approximately 11,000 Duna speakers. Ipili, Enga, Huli, and Nete are all in the Trans–New Guinea Engan language family. Hewa is from the Sepik Hills language family. Duna is an isolate in the Trans–New Guinea phylum. Culturally the Ipili, Enga, and Huli are quite similar, with intensive mounded agriculture oriented toward pig rearing. The Duna and the Nete also practice a similar subsistence pattern but are feared in Porgera for their knowledge of sorcery and black magic. The Hewa are foragers who also practice nonintensive shifting cultivation. In Porgeran myths Hewa are always demons and are greatly feared even today for their knowledge of the dark arts.

This influx of outsiders is due to Porgera's reputation as "a money place." Landowners currently hold a 2.5 percent equity share in the mine and also receive quarterly royalty checks worth around 1 percent of the mine's output. Annual production has averaged over 750,000 ounces of gold since the mine began, and between 1990 and 2009 a total of K1.18 billion (approximately US$600 million) has been paid to Porgerans in the form of wages, royalties, and compensation for lost land (Barrick Gold Inc. 2010). In 1999 I surveyed forty households in nonlandowner communities in Tipinini and found that the average monthly income was

K62 (about US$20). A 1992 survey (n = 35) corroborated these numbers, indicating that nonmining compensation amounted to approximately K70 per month (Banks 1999: 113, table 3.8). While many imagine that Porgera is a money place, in reality it is not so for most. Like most of the resource frontiers around the world, there are incredible inequalities in the livelihoods of frontier dwellers. In the following chapter I examine the historical processes that led to the creation of Porgera's resource frontier.

*Historical process of resource frontier & how
& what inequalities emerged as a result.*

CHAPTER 2

Colonialism, Mining,
and Missionization

Breaking Down the Fence

Interviewing an Elder about First Contact, May 1999

Jerry: Pesaip, what can you tell me about the time that whites first entered Porgera?

Pesaip: There isn't much to tell; they killed some pigs, raped some women, and shot some men. (Rises slowly and cackles as he shuffles off to his house)

In 1926 news of a gold strike along Edie Creek on the northeast coast of New Guinea reached Australia. Initial reports were that teams of miners were "winning up to 200 ounces of gold a day" (Leahy 1991: 3). A gold rush ensued as Australians poured into the region in search of instant riches. By 1930 miners were in search of "another Edie Creek" (6), and the Australian prospector Michael "Mick" Leahy led an expedition into the highlands of New Guinea, which on maps of that time was merely marked "unexplored." Leahy's travels into the highlands from 1930 to 1934 have become legendary in anthropology through the film and book *First Contact* (Connolly and Anderson 1987). Although the Australian colonial administration thought the highlands were uninhabited, the expeditions found hundreds of thousands of uncontacted people living there. In March 1933 Leahy, his brother Dan, Ken Spinks, another prospector, and an administration patrol officer named Jim Taylor led an expedition into the Wahgi valley, where Mt. Hagen is located. Along with the four white men were dozens of carriers, native police, and other camp followers. Upon their return the *Sydney Morning Herald* of November 16, 1933, announced, "Stone Age natives discovered in New Guinea." The expedition was reported to have "covered 1000 miles" over five months, "traversed an area of 4500 square miles, and came in contact with 200,000 natives whose existence was previously unknown."

The following year, 1934, two separate prospecting expeditions pushed east from Mt. Hagen into present-day Enga Province. The first expedition was Mick and Dan Leahy's abortive attempt to prospect as far as the Dutch border. As they entered the area, they soon realized that the Enga perceived the prospectors in different terms than had the groups to the east. Bob Connolly and Robin Anderson (1987: 197–98) describe their expedition as "a grim, bloodstained journey. . . . The people generally thought the strangers were spirits who had come from the sky, but spirits who were not necessarily either feared or respected [although among the Ipili fear was one of their responses; see below]. The response was neither tearful welcome of the returning dead, nor abject terror, but confusion. This made the first encounters unpredictable, and therefore much more difficult and dangerous for everyone concerned, black and white." Just west of Wabag, the Leahys feared attack from a massed group of warriors, and Mick and the native police shot into them, killing some fifteen people (199).

In the second expedition into Enga, two twin brothers, Jack and Tom Fox, left Mt. Hagen on an illegal prospecting trip that intended to succeed where the Leahys had failed. Since the Foxes had no administration approval for their trip they traveled light and fast, with just a few armed native "police," and prospected along every stream they crossed. They were unsure of their exact route (Fox and Fox 1936) yet believed that they had made it to the Dutch border and had failed to find a trace of gold in any stream. Later expeditions found gold downstream of the Porgera River along beaches of the Lagaip River, so it remains a mystery why the Fox brothers never did. One answer may have been the hostilities they provoked along their path, which prevented them from spending sufficient time prospecting. Typically prospectors obtained food from trading with local people. With no trade goods or supplies for their "monthslong jaunt,"[1] the Fox brothers would have extorted food from people with threats of violence. Kainjem Yawarin of Wailya recalls their coming: "I was a little girl when the whites first came. The first two that came [the

Foxes], they came with guns and just took our pigs without giving us anything. Later, when whites came, they traded shells and salt with us for our pigs. It was the police of the white men that were the worst, they would steal and kill our pigs" (field notes, June 1999). A man from Wailya named Kaiyamba had seen the Fox brothers at Mulitaka and realized they were coming toward Wailya. He hurried back to Wailya and told everyone to prepare to attack the whites when they arrived (see Jacka 2002 for more details). Before Kaiyamba could attack, the Fox brothers killed him and another man, named Lone. Whereas the main trail out of Wailya should have taken the Fox brothers into Porgera, based on the lack of oral histories about them in Porgera, it appears that they continued down the Lagaip River valley, missing Porgera's gold. During their expedition the Foxes killed unknown numbers of people. However, in the Tari Basin among the Huli, they are reported to have killed forty-five people in twelve days (Allen and Frankel 1991a).

The killings by the Leahys and the Foxes were eventually excused by the administration authorities, who deemed violence against highlanders necessary for preventing the bloodshed of white Australians. However, with the killings of two European Catholic missionaries east of Mt. Hagen by local people, the administration decided to declare the highlands an "uncontrolled" area in 1935, preventing any further encroachment into the region by any outsiders who were not already there (Connolly and Anderson 1987: 216). The prospectors and missionaries who stayed on in the uncontrolled areas were confined to the areas around their camps. In effect, this closed the PNG highlands to outside contact until after World War II.

There was to be one exception: the Mt. Hagen–Sepik Patrol. Despite the failures of the Fox brothers to find gold in the western highlands region of the Territory of New Guinea, there were intense external pressures to explore this area and open it to the outside world. Oil and gold prospectors believed there were major resource fields in the region. Both Oxford University and the British Royal Geographic Society proposed scientific expeditions to explore west from Mt. Hagen; however, the administrator of the territory, Walter R. McNicoll, refused all applications until administrative control could be established. Finally the minister for territories, Sir George Pearce, urged McNicoll in late August 1937 to consider bringing the area under control for the purpose of allowing the entry of oil prospectors (Gammage 1998: 17). At the end of 1937 McNicoll decided to mount a patrol west from Mt. Hagen. The objectives for the proposed Hagen-Sepik patrol were to look for gold between Mt. Hagen and the Dutch border, to establish friendly relations with the native populations, and to find a base from which to "administer native populations and supervise the activities of miners and others when the area ... is opened for the entry of European expeditions and parties prospecting for oil, gold, etc."[2]

The Mt. Hagen–Sepik Patrol

The patrol lasted from March 1938 to June 1939 and was led by Jim Taylor, who was the first patrol officer (kiap in Tok Pisin) to set up an administrative center (called a patrol post) in the highlands region in 1932. His second in command was John Black, a former highlands patrol officer and now the assistant district officer in the eastern highlands. They were joined by an Australian medical orderly, Pat Walsh, 233 carriers, and several native police, making for an expedition of 271 people in all (Gammage 1998: 48). At the beginning of his diary documenting the patrol, John Black wrote, "It appears that mining interests have forced the issue [of a government patrol]. The Administration has been forced to attempt to open the closed book of its uncontrolled areas."[3] Over seventeen months the Hagen-Sepik patrol covered 3,000 kilometers through the western highlands of New Guinea Territory, passing through Porgera three times. On the second trip, in 1939, they discovered gold in the Porgera River, and as a result the patrol spent over a month prospecting and interacting with the Ipili.

I focus on the Hagen-Sepik patrol because it initiated the realization of the opportunities that Porgerans and others were able to capitalize on by attaching themselves to the patrol in the guise of interpreters and guides. Many of the young men and boys who joined the patrol eventually became some of the key figures in promoting the development of Porgera in the postcontact years. Not only did the Australian colonial government see the potential wealth of this area as beneficial for the metropole, but many highlanders saw the patrol, and later the colonial government, as a source of power that could be exploited. Men who were taken on by the patrol were sometimes unclear initially about their status within it; many of them described being "jailed" or "held prisoner" when recounting their experiences. Pongolo Yawe told me, "When the whites first came they arrested me. They held six of us, Tongope, Kuala, Ipape, Igili, Ekondale, and me. I ran away in the night." Sixty years later Pongolo was relating this story somewhat wistfully, as Tongope, Kuala, and Ipape all became rich, influential leaders in the early mining years. Another reason I focus on this patrol is that the violent encounters of first contact shaped people's attitudes about the whites in profound ways that would resonate for several more decades. The histories of first contact also demonstrate that the frontier was not a place of lawlessness but that retribution and killings were a central, if implicit, policy of the colonial venture (Banivanua-mar 2007; Kituai 1998).

By the time the patrol reached Porgera in September 1938 it had already split into three groups as it was too large and unwieldy; a patrol that size could not enter an area without eradicating local food supplies. Additionally policing the line of carriers was difficult. Several months of the patrol passed before John Black realized that carriers (and even the police) were raping local women and stealing pigs for personal consumption. On September 10, 1938, Taylor and his group arrived in the Wailya area after following the Lagaip River downstream. In 1999 in Wailya, Lekiolo Kamino remembered the coming of the whites, noting that they had come four times in fairly close succession. (The Hagen-Sepik patrol passed through three times; the Fox brothers were the first of these four.) Lekiolo recalled, "The first time the whites came through, they passed by quickly. They just shot our pigs and left with them [Fox brothers]. The second time they came they killed our pigs too, but this time they paid us with steel axes. Tuyepe, Malo, and I each got an axe for having our pigs taken."

That night Taylor slept at Polokombi and the next day at Wailya (what Taylor calls Wala and Black calls Walye in their journals). When Black came through several months later, people complained that Taylor's patrol had "played little hell—shot all pigs" and raped a number of women. On September 11 in Wailya, Taylor (MS 9218/5/35) writes in his journal, "We are in the heart of the limestone country," and worries that the possibility of gold discoveries is lessened by the fact that the rivers start in limestone country. Little did he know that a few miles to his south was the source of Porgera's gold in Mt. Watukati. Taylor also starts to appreciate the high country of Enga and lists a number of development opportunities that one could pursue "if one can but see it," such as dairying, tea planting, gold mining, pig raising, and coffee growing. The following day he intended to continue down the Lagaip valley, but local guides insisted that there were no settlements in that direction. (There are none.) His guides instead led him up over the immense Mt. Auwakome to Lese (Leze in Taylor's journal), finally arriving in the Porgera valley on September 12.

As the patrol rested for a few days in Lese, Taylor took photos of the mountainous panorama of Porgera. He took several shots of Mt. Asienda (what he calls Kaijenda), a "beautiful but terrible" mountain (figure 2.1). He writes that the people are "very decent" and similar to the Huli. He

Figure 2.1. Mt. Asienda from Tipinini, Lakeame's waterfall on the right.

notes later that they are "decent and friendly showing a real interest. Not just playing the fool as some new people do, who regard the white man as an object of merriment and butt for their jokes" (MS 9218/5/37). Over the next few days Taylor's patrol moves across the Porgera valley and into the Paiela valley. He notes that most of the Porgerans act as if they are handling steel axes and knives for the first time. While Taylor's visions of the development of the highlands would eventuate in his pursuit of these endeavors in Porgera and the eastern highlands, Black's experiences in Telefomin prompted a very different attitude to the area. By his second diary of the patrol, Black expounds at length on his thoughts of the natives and how administration policies should treat them. He writes that the natives are not child-like but "shrewd" and pay "lip service" to whites they see as powerless and "play off" white officials, civilians, and missionaries against one another. The people with real power were the police, in Black's opinion: "Police soon rank higher in the mind of the uncontrolled new native than the whiteman because they observe that the former have unlimited privileges: women, power, pork—while the whiteman's for some reason is strictly limited."

In early 1939 Black's contingent began the trek back to Wabag. Since leaving Telefomin, Hewa had been attacking Black's patrol using guerrilla tactics, and Black and the police killed several Hewa in response. On March 13 they entered the Paiela valley among Ipili speakers, and the attacks ceased. People were also willing to trade food with the patrol for the first time since their leaving Telefomin. Black writes, "Natives friendly as usual and accompany us singing and dancing as we go." At Kolombi on March 18 he reported being called "Yelyegurri. The white man from the skies." Yalyakali is the Enga word for "sky man"; Ipili speakers would have said tawe akali. (Black's translator, Yaka, was an Engan.) Later that same day Black's attitude started to change, as he feared an attack: "Taylor thrashed an old man here on his first visit but shot neither pigs or men. The locals might want to back this insult in dawn attack." The police told Black that the Ipili were trying to use "sleep magic" on the patrol so they could raid them. The following morning he wrote, "Ponder on my assessment of local native mental attitude to ourselves. I think I am right that they are inclined to be truculent and that we will have to be keenly alive to hostility from now on. If Taylor had had to shoot someone in this valley it would have made travelling much more safe for ourselves and for future travellers. If we meet with no resistance justifying the opening of fire ourselves not shooting simply means putting off the evil hour and will ultimately mean the loss of more lives than would be the case with prompt and bold action on the first occasion of truculence or refusal to disarm."

Yaka tried to tell Black that the Paiela Ipili were not hostile, merely afraid as they thought that Black came from the sky world (see chapter 3). The discrepancy between what Black thought people were doing and what the Paiela Ipili thought was occurring illustrates the potential for misunderstandings to develop in these initial encounters. The Porgera valley Ipili indicated to me the same levels of confusion when they encountered whites, and not just in initial encounters but even into the late 1940s. People recalled that they were unclear where the outsiders had come from and what they even were—sky people or demons.

On March 25, 1939, Black and the patrol entered the Porgera valley. While crossing the Porgera River, Black finally found what countless others had dreamed of: gold. Later that evening he recorded in his diary, "Gold +ve [positive] big rough edged flat colours in every dish. Best prospect I have ever seen let alone found." The rough-edged flakes indicated that the source was very near, not smooth, water-worn gold from afar (Gammage 1998: 188). As best Black could tell, the source of the gold was the upper Porgera valley, which held an extensive population. He indicated in his diary that the large population would make the search for gold more difficult. Also the Porgerans were being "truculent" and "obstreperous." When the patrol tried to trade for food, the men would lift their leaf buttock covers—"unmistakable signs" that the patrol could forget about eating sweet potatoes. When some of the young men were being loud and rude, Black "cracked" a few over the head with his rifle butt. He lectured the elders about their "unpleasant attitude" to the patrol and then cracked a few more heads. He justified this "demonstration" by noting that since most Porgerans wore thick wigs of human hair decorated with bird feathers and leaves there would be no damage to their anatomy. Afterward people brought in two pigs, and a few of the young men indicated that they "want[ed] to join my circus and explore civilisation."

Black continued to prospect for about a week in the upper Porgera. On March 31 a plane flew over and later an old man asked Black "if I and the aircraft are devils, spirits or what. He is more than inclined to doubt my denial." The following day Black and the patrol set out to the east to return to Wabag. In the eastern Porgera valley at Ole (Horei in the patrol report), Black had a long talk with an old man who wanted to know if "we were spirits or what. My white skin and clothes he says he had never seen before." At Ole, a man named Kuala (Gwara in the official Hagen-Sepik Patrol Report) joined the patrol. Kuala stayed on with Taylor after the patrol and eventually was one of his head "boss-boys" when Taylor set up an alluvial gold operation in Porgera in 1948; as mentioned, he later became a major player in alluvial mining in the late 1950s and 1960s. Years later Kuala reported that Yaka, Black's Enga interpreter, had "no sense . . . raping women, beating people up, killing pigs" (Gammage 1998: 189).

With Black's patrol in the valley a mere nine days, the Porgerans decided that they had had enough of the rape and abuse. Elders I spoke to in 1998 and 1999 told me that Pakasia, a leader living in Lese, decided he was going to attack the patrol and kill the white man in charge of it. Pakasia came to Black's camp at Lese on April 3, and, armed with a spear, he made threatening gestures to the camp and the sentries. He then made a taboo mark on the trail leading to the hamlet. He had come the day before as well and made a "military assessment" of the camp, according to Black. Pakasia and other Pakoa dug a trench alongside the trail and put a barricade along it so that they could shoot arrows at the patrol as it came abreast of their fortifications. Black, however, told his police to seize Pakasia, which they did, and then had him flogged with a rattan cane. After being released, Pakasia ran to get his bow and arrows and join the others in the trench. Yaka, Black's interpreter, warned that an attack was imminent, so Black ordered the police to open fire on the men in the trench. Pakasia was killed immediately, as were two others, Londe and Mipu. Another man, Katapene, was shot in the heel. After shooting him, Black preceded to stitch him up. "About 6 stitches—nasty wound—bone of heel smashed. I think he will recover," Black wrote in his diary.

Minala Nolapa and Pes Bope were both young men when Black was at Lese. Minala recalled, "We heard rumors that white men were coming. We didn't know what they were as they had so many strange things and they were very smelly. We were afraid of them and wanted to kill them. The men who came as carriers and police took our women and raped them and chased the men away who were trying to defend our women. We didn't want these men around, so we tried to chase them away. We were formed in a line and then they shot into us. As they shot us, bones, stomachs, and intestines went everywhere and we thought these people were *yama* [demons] because there were body parts everywhere" (field notes, December 1998). Pes remembers, "We were hiding in the forest, the women, children, and pigs, because we were afraid of the white men. Then we heard guns shooting. We didn't know what it was and we thought the world was ending. Then we found out that four men had been shot by the white men. Three of them died. Everyone was saying that the whites were yama with powerful spears who were going to eat us. Later, as the patrol climbed up Mt. Auwakome toward Wailya, they were mocking the cries of the mourners at Lese, saying iiiyiiyiiyii, iiiyiiyiiyii" (field notes, February 1999). Afterward Black wrote of the climb up Mt. Auwakome, "Shortly after our departure the voices of mourners sadly lifted up to us through the mist. I feel sorry for these fine mountaineers and then I realize if strong action had not been taken it is we that would have been mourning for our own slain." He justified the shootings: "I am glad the ringleader was shot and that shooting was accurate. It should save many lives in this valley and stop future hostility to travellers."

When Black arrived in Wabag, Taylor was ecstatic to hear the news of the gold find. He convinced McNicoll to allow him and Black to explore it further (Taylor MS 9218/5/46). While Taylor went to Tari to return some Huli carriers to their homes, Black returned to Porgera for another two weeks of prospecting. Along with Porti, the patrol's main guide, the two found traces of gold in the Kogai River, leading Porti to comment, "It is like another Edie Creek." Black and Porti narrowed down the source of the gold to somewhere in the mountains between the hamlets of Yairik and Mungalep. Black speculated that the find was sizable as he started to find pyrite and galena, indicating "a good gold bearing reef somewhere above us or at least possibility of one." By that time Black realized that much of Porgera's alluvial gold was contained in the Kogai, a narrow, boulder-choked gorge that would be difficult to work. Porti meanwhile insisted that there was payable gold in the wash. Taylor arrived several days later, and he and Black decided, "Gold is not a big show. But it might" (MS 9218/5/47a). In his journal Black writes that Taylor will talk of the need for prospecting the Sepik headwaters and "will recommend it in his report." Taylor and Black had other plans, though, for in the official report Taylor (1939: 353) writes, "Black had examined the rivers where he found gold, very closely, but they were nearly valueless and would not be worth developing in the future." In the official patrol report, Taylor even wrote on the first page of the document, "No great discovery of a practical value, at any rate none which could be put to immediate use, was made except perhaps in a negative sense. Many people, hoping that it will be so, say that they expect great goldfields to be discovered and developed in the western half of Australian New Guinea. I hoped so too, because the enrichment of the Territory will be of the utmost benefit to the inhabitants. . . . The Government would be ill-advised however to harbour any such illusion. Plans for our future should be based on an hypothesis of agricultural development."

The Hagen-Sepik patrol was the longest, most expensive patrol ever conducted in what is today Papua New Guinea (Gammage 1998: 211). Given that Australia had declared war on Germany on September 3, 1939, and that Japanese military were already in Rabaul posing as fish researchers and merchant seamen, the Hagen-Sepik patrol report was suppressed by the government for fear that it would be used to gain intelligence on locating airbases in the highlands (214). As a result few people outside of Papua and New Guinea ever heard of the Hagen-Sepik patrol. With the onset of World War II, it would be almost a decade before whites returned to Porgera.

The Impact of the Sky People

Taylor and Black's arrival in Porgera caused much confusion. People were initially unsure what category of being they were. Were they demons (yama) or sky people (tawe wandakali)? As discussed in more detail in chapter 3, sky people are generally benevolent yet aloof figures in Porgeran culture. They are responsible for rain, thunder, and lightning and have the ability to bestow wealth and beauty. Socially transgressive behavior in the high-altitude rain forest brings their wrath and punishment. The land of the sky people is always described in stories as a place of wealth and plenty, with larders full of wild game and garden produce, where the people are happy, healthy, young, and beautiful. The patrols were well stocked with trade goods—steel axes, face paint, and shells—that were highly coveted by Porgerans. With seemingly unlimited wealth and power in the form of guns, Taylor and Black came to be perceived as sky people (Gammage 1998). The violence that marked their arrival in the valley, their abrupt departure, and a series of climatic, social, and ecological calamities that ensued after they left eventuated into a series of ritual attempts to either return things to the status quo or to gain access to the sky world, depending on where the rituals occurred.

Not only did the Hagen-Sepik patrol bring death from gunfire to Porgera and the western Enga area, but the police and carriers also spread gonorrhea (Gammage 1998: 200). In 1940 and 1941, just after the patrol's departure, severe droughts and abnormally heavy frosts devastated the gardens throughout the high country of western Enga (Ballard 1995: 88–89; Meggitt 1956: 92). Then in 1942 heavy rains flooded the gardens in the region, causing widespread famine (Meggitt 1956: 92). A major influenza epidemic hit the western Enga area in 1943.[4] The following year a dysentery epidemic started in Mt. Hagen and reached eastern Enga in October 1944; by November dysentery was raging in Wabag and making its way farther west (Wabag PR #3 1944/45; Wabag PR #4 1944/45). Later patrol reports (Wabag PR #2 1947/48; Wabag PR #3 1947/48) indicate that the dysentery epidemic extended "far and wide," impacting areas well to the west of Wabag, and there is even mention of a small dysentery hospital set up in Sirunki, west of Wabag. In some parts of the highlands, between one in six and one in eight people died from the combination of frosts and flu and dysentery epidemics (Gammage 1998: 230). In addition to human casualties, pigs also fell victim to several debilitating epidemics; one of the most significant was a "swine fever" that came through the area in August 1944, killing around 80 percent of all the pigs around Mt. Hagen (Hagen PR #1 1945/46). This disease then swept to the west, although killing fewer pigs than at Mt. Hagen (Meggitt 1956: 92). One patrol officer noted that there were two waves of pig epidemics; one was a neck swelling disease with 100 percent mortality, while the other was a pneumonia with about 50 percent mortality (Hagen PR #1 1945/46). Subsequent researchers have identified these as porcine anthrax and bacterial pneumonia (Meggitt 1973; Ballard 1995).

In early 1942 the Japanese seized the islands and coastal areas of northern New Guinea. In the highlands Australian New Guinea Administrative Unit (ANGAU) officers oversaw patrol bases to help alert the Allies to Japanese attempts to come overland and attack Australia from the north. Spotting stations were set up across the highlands and staffed by members of the New Guinea Air Warning Wireless Company (Gammage 1998). Wabag received its first spotters in November 1942. In September

1943 Capt. John F. Clark (1944) arrived at the Wabag post to watch for the Japanese and help any Australian Military Force personnel escaping from the northern side of the island. Clark reported little military activity during his year-long sojourn there, yet it was hardly an uneventful time, given the human and porcine epidemics that were raging. With wartime rationing, the spotters in the posts in Enga took food from locals, yet failed to reciprocate. The diseases, climatic anomalies, and social disruptions prompted the Enga to attack spotting posts at Sirunki, Birip, and Pumunda, culminating in a massive reprisal slaughter at Pumunda by Clark and his police (Gammage 1998: 230).

In western Enga around Porgera there were no whites present to whom the people could direct their anger and blame for the disasters that had befallen them and turned instead to ritual means. In late 1943 Dan Leahy entered western Enga, traveling to Tari from Wabag along the headwaters of the Lai (Purari) River, thence returning via the Wage and Lagaip rivers. Despite people starving along the whole route from the previous frosts, droughts, and rains, Leahy's patrol was warmly welcomed (Meggitt 1956: 92). Within two years, however, numerous groups throughout the region were participating in a "pig-killing cult" (92). In late 1945 ANGAU Capt. Neptune Blood traveled from Wabag to Tari, "investigated the recent rapid spread" of the cult, and "tried, with some success, to dissuade the natives from maintaining it" (92).

The "pig-killing cults" among the Ipili, western Enga, and Huli in the early 1940s have been analyzed in detail by me and others (Biersack 1995b, 1996b, 1998b, 2011a, 2011b; Feil 1983; Frankel 1986; Gibbs 1977; Goldman 1983; Jacka 2002; Meggitt 1973; Sharp 1990; Wiessner and Tumu 1998, 2001). I will provide just a brief overview here. The cults were a novel response to the series of social and ecological disturbances that traditional ritual activities were unable to set right. One of the key messages in the cults was to destroy (kill pigs and break shell wealth) customary forms of wealth in preparation for new kinds of wealth that were to come. In and around the eastern Porgera valley, cult leaders also stressed that the rituals would transport everyone up to the land of the sky people to live immortal, labor-free lives. In the Tipinini area in eastern Porgera, two cult leaders, Ipayema and Timbapu, who prophesied that the cult practitioners would ascend to the sky world, led a small group of practitioners into the high-altitude rain forest and killed them as they peered into a lake that was to transport them to the sky world. A few people survived and informed others of the events, and one of the leaders was killed and one driven off. Another key message in the cults was to give up several traditional practices, such as warfare, gender segregation, ancestral sacrifice, and fears of female pollution. Special platforms were built for killing pigs, and sacrificed pigs would be held up and offered to the sun instead of to the ground, where the ancestors lived. According to interviews with some of the participants in the cults around Porgera (see Jacka 2002), a portion of the cults' proscriptions came from men who had joined parts of the Hagen-Sepik patrol. Therefore the cults were not just novel attempts to deal with frosts, sickness, and death but also a means to incorporate the coming of whites into local frameworks of understanding (cf. Sahlins 1981; Schieffelin 1991).

Within a short while most people stopped participating in the cults as no new forms of wealth arrived. Today, despite the cults' initial failure, most Porgerans I have spoken with consider the main leader of the cults, a man named Wambilipa, a "prophet" of Christianity. New forms of wealth and new social practices did eventually come to the area, not just in the six months that Wambilipa had prophesied. People also claim that the beginning of the cults in the early 1940s was contemporaneous with the crucifixion of Jesus. Jesus's death sent a vision to Wambilipa to "open the roads" for colonialism and Christianity. Indeed, with the end of World War II, colonialism and Christianity soon arrived in Porgera.

Colonialism in Porgera after World War II

Leaders, called "big men," in societies across the highlands depend on persuasion to influence their clan members (Sahlins 1963; Strathern 1971). As a consequence, when the Australian colonial administration arrived in the highlands, it had to create political and legal systems with which to control the people (Kituai 1998). The systems that were created depended on initiating new forms of political leadership, setting up village courts, and establishing administrative centers headed by colonial patrol officers called *kiaps*. Patrol officers were "responsible for discharging Australia's main obligation to the League of Nations, to civilise the people of New Guinea"; as the core of the colonial administration, "with powers and prestige above all other departments," they "could combine swashbuckling adventures with innovative policies" (Gammage 1998: 9). To create political leadership in the near absence of organized forms of authority, patrol officers would appoint individuals with the aptitude for leadership, usually the big men, as *luluais* and *tultuls*. The luluai was the paramount authority in the hamlet, while the tultul was his lieutenant. After the patrol officers had familiarized the local people with court proceedings (in which the patrol officer was the judge, jury, and penal force), village court magistrates were also appointed. Both of these systems, the political and the legal, were established only after fairly long periods of interaction between the colonial administration and the people. Some thirty years passed between first contact with the Porgerans and the time that luluais, tultuls, and village court magistrates were finally appointed.

The establishment of administrative centers, called patrol posts, developed over time as successive areas fell into the ambit of colonizing and "civilizing" practices. The greatest distinction was between areas considered "controlled" and "uncontrolled." The typical pattern in the Enga region was that with the initial establishment of the patrol post at Wabag, the surrounding regions would be patrolled and controlled in an ever-increasing compass. Once an area had been in frequent contact with a patrol officer and was pacified, a new patrol post would be set up in the region, and the shift from uncontrolled to controlled area would be accomplished. Once control had been established, missionaries and other whites would be allowed in the region to further complete the process of colonization and incorporation into the colonial society. The colonial administration expected that as little bloodshed as possible was to occur during the pacification and colonization of the highlands. However, highlanders resented the intrusions into their lands, especially the activities of the police and carriers who accompanied the white patrol officers. The very structure of the patrols made it impossible for them to peacefully work their way across the highlands, and as a result much bloodshed ensued (see Kituai 1998).

After World War II the prospectors and patrol officers in the highlands turned their attention back to Porgera. Taylor, Black, the Leahy brothers, and a few other patrol officers formed a company to begin mining Porgera's gold but were denied permits by the colonial administration. This failed to prevent their entry into Porgera. Nonetheless the partners quickly realized that the area was too remote and rugged to make it a worthwhile venture.[5] Joe Searson, one of the partners, decided to try it alone and received a permit for alluvial mining. With the permit granted, the administration decided to establish a patrol post at Porgera to ensure Searson's safety (Wabag PR #3, 1947/48). Between May 1948 and March 1949 hundreds of "storepedoes" were dropped by plane to set up and supply the Porgera patrol post, Searson's operations, an exploring party led by Bulolo Gold Dredging, and one other potential expatriate prospector. In addition to food and medical supplies, the storepedoes contained axes, machetes, knives, mirrors, salt, buttons, matches, and other trade goods. The same night that one drop of twenty-two storepedoes was made, the patrol officer reported that the people stayed up all night "singing" at the drop site (Wabag PR #2, 1948/49), which suggests that the sky world cults were efficacious. In particular people desired steel, trading a pig for one axe head: "Now that much steel has been introduced into the area large tracts are being cleared and new gardens being prepared" (Wabag PR #3, 1948/49). One of the women's favored trade items was not what the storepedoes contained but rather the parachute material, which they would unravel and use in place of plant fibers to weave net bags.

By March 1949 only Searson remained in Porgera, as the other hopeful prospectors considered the gold too difficult to mine. In response the administration decided to close the Porgera patrol post and conduct patrols out of Wabag and Laiagam annually or biannually. Searson employed about forty workers who ran four sluice boxes that provided an average of two ounces of gold per day (Wabag PR #6, 1948/49). In September 1950 a prospector named Jim Brugh started working Searson's second lease along the Kaiya River (a tributary of the Porgera). Relations between Searson and Brugh and the laborers had deteriorated somewhat by this time (Wabag PR #4, 1950/51). Searson was out of the valley, and some of his workers had stolen his tools and were mining gold on their own. Food for the workers was in short supply, and several of them complained of not getting paid for their work. The patrol officer noted, "There is a lot of gold in the Pogera [sic] and the miners appear to do quite well. They should be prepared to spend some of the returns on the equipment of their labour" (Wabag PR #4, 1950/51). Between 1949, when the patrol post was closed, and 1954 only six patrols came through Porgera. With the exception of the area around Searson's and Brugh's camps, most Porgerans hid from the few patrols that entered the valley, especially in the eastern Porgera, where Black had shot the men. In 1954 the patrol officer complained, "For 7 years now casual patrols have been made through this area. The effect on and benefit to the local peoples appears to have been nil. They are frightened and distrustful of us. We know little or nothing of them" (Laiagam PR #1, 1954/55).

During the 1950s the only information to come out of Porgera in the patrol reports pertained to tribal fighting, what one patrol officer called "a semi-permanent feature of the area" (Laiagam PR #3, 1955/56) and another "a continual source of annoyance" (Laiagam PR #2, 1956/57). However, by late 1957 Wabag Assistant District Officer Macilwain was "of the opinion that the native situation [i.e., tribal fighting] has considerably improved since . . . [1948 due] to the fact that the leaders have accumulated wealth from the mining activity, and that they are reluctant to lose this wealth as a result of fighting" (Wabag PR #2, 1957/58). An important event happened to decrease fighting in the area: Porgerans themselves began mining rather than working as laborers. Claims were also transferred among the expatriates. In 1955 Searson turned one of his leases over to Jim Taylor, and in 1957 Brugh handed over Searson's second lease to Taylor. From 1955 until he sold his leases in 1980, Taylor maintained mining from afar, coming to Porgera only a few times a year to pick up gold and pay his workers, who were overseen by Tongope, another young man who had joined Taylor on the Hagen-Sepik Patrol in 1939.

Porgeran Miners

Sometime in 1956 or 1957 Puluku Poke, a former native medical orderly in the Porgera region, began mining along the lower Porgera River, just downstream from Taylor's lease, and became the first indigenous miner in the region. Pawe Lembopa, a native medical orderly stationed at Tipinini, started his own operation some six to eighteen months later, along the Kaiya River, a tributary of the Porgera. Similar to the system set up by Searson, Brugh, and Taylor, Puluku and Pawe hired bosbois to oversee the labor and then also hired laborers to work their claims. The bosbois earned between 2 and 5 Australian pounds per month, while the latter earned about 1 pound, 10 shillings per month (Laiagam PR #2, 1959/60). At about the time that Puluku and Pawe were given official leases (May 1959), two other Ipili men, Akiwa and Waliya, attempted to claim leases on land that Puluku had already applied for. This would be the beginning of a long series of struggles by Porgerans to obtain lucrative alluvial mining lands.

Much of the tension over alluvial mining stemmed from conflicts over group-owned resources and individual ownership. One patrol officer wrote, "Another visit by a Mining Assistant appears to be a necessity and should be arranged as soon as possible. Many of the locals are interested in working gold but have been warned not to infringe upon the pegged claims. They, however, regard anything within their land boundaries as belonging to them and cannot see why gold recovered within those boundaries is not theirs. It is hoped that a mining assistant may be able to persuade one or two of the entrepeneurs [sic] to reduce the size of their claims to allow of others finding gold" (Laiagam PR #2, 1959/60). By May 1960 "the native mining situation was getting out of hand. . . . [It] has degenerated into a seething mess of intrigue in which there have been threats against the life of native miner Puruk [Puluku]" (Laiagam PR #4, 1959/60). When the patrol officer, Graeme Hogg, entered the alluvial mining area in May 1960, he noted that he was immediately "besieged"

Figure 2.2. Contemporary alluvial miners who have breached the tailings waste pipe.

by requests from six men to also be allowed to mine for alluvial gold. Hogg wrote that they were actually interested in mining only on Puluku's claim as "Puruk's recently acquired wealth is a source of very obvious jealousy, which combined with the local attitude that the landowners are in the right, has developed into an extremely unhealthy situation." Much of the resentment came from the fact that Puluku was mining on land that belonged to his wife's clan, not his own. Prior to leaving, Hogg promised to have a mining warden visit the area to issue leases to the men who were interested in pegging their own claims.

All the patrol officers in Porgera agreed that despite the biannual patrols through the region, the Ipili had very little regard for colonial laws and policies (figure 2.2). Hogg writes, "The only conclusion to be drawn from this report is that these people must be brought under much more effective control in the near future. The local inhabitants do not appear to be the type who can be brought under influence by passing patrols, however frequent." B. McBride, in the region in July 1960, noted that Pawe and others had begun to encroach upon Taylor's mining lease, and concluded, "I consider that a GOLD CONSCIOUS IPILI people whom [sic] have little respect for any of our laws at present will work the lease whenever they think they can get away with it" (Laiagam PR #1, 1960/61).

To dispel some of the tensions among the Ipili miners, McBride's patrol included an official from the Division of Mines, N. Robinson, to stake more claims for some of the disenchanted landowners. Puluku's original claim, which extended for 1,680 meters along the Porgera River from its confluence with the Pongema River, was broken up into four claims. The initial claim had included lands far outside even his wife's clan, and with such a large claim, Puluku had been unable to prevent others from working gold on his claim. McBride wrote that the chief offender in regard to claim-jumping

> was without doubt the ex Administration Interpreter Kwela [Kuala]; a man taken out of the Porgera Pre-War to Rabaul [by Jim Taylor in 1939] and later employed for years by the Administration in various parts of the Highlands. Kwela openly admitted that he was the main instigator encouraging others to work their clan lands on Puruk's lease. Kwela's and Puruk's clans are traditional enemies and part of the lease extended into Kwela's clan boundaries. Kwela, as the spokesman for his and other clans expressed the view that the people considered that the new found wealth was not being distributed evenly and that Puruk was winning gold from land that was not traditionally his.

In addition to problems in the lower Porgera between Puluku and the other claimants to mining land, Pawe's claim-jumping of Taylor's lease in the upper Porgera indicated that claims in that region needed to be established as well. As such, Robinson also staked out seven claims for landowners in the upper Porgera along Yakatobari (a.k.a. Yakatipati) Creek.

The colonial administration had a great interest in promoting indigenous mining so as to bring the Ipili into the "civilized" world. From July 26 to August 6, 1960, nine Ipili miners were taken on a tour of Goroka, Kainantu, Wau, Bululo, and Lae. The purpose of the trip "was to enlighten these primitive people and allow them to observe mining activities practised by both Europeans and Natives in other centres, and also to show the economic development that can be attained by individual enterprise" (Robinson 1960: 1). Patrol officers familiar with the Ipili had a different impression of them, however, as B. McBride (Laiagam PR #1, 1960/61) reports: "These people without doubt know our main laws and are familiar with our ways. They are by no means newly contacted primitives but killings and clan skirmishes appear to be the order of the day; my impressions are that the native situation has deteriorated instead of improved over the past years. Other officers have reported, and I agree, that familiarity with our law has brought contempt and disregard instead of respect." The immediate impact that the tour had on the nine Ipili men was uncertain; in one man's case it was noted that a "remarkable change" had occurred as he was now intent on beginning mining in earnest, while another man "had no sooner stepped from the aircraft when he seriously assaulted his '*meri*' [wife] with a piece of timber and was sentenced to two months prison detention" (Robinson 1960: 3).

To promote economic development the administration set procedures and policies as to how the gold won by the alluvial miners would be sold. Every three months an officer of the Commonwealth Trading Bank at Mt. Hagen would travel to Laiagam to purchase the gold from the Ipili miners. Money would be paid immediately to the miners, which they could put into a savings account or spend. Many chose to put some of their earnings into accounts, which led one patrol officer to exclaim, "Various patrol officers stationed at Laiagam have encouraged indigenous enterprise over the past few years resulting in the strange situation of natives classified as 'uncontrolled' possessing bank books with sizeable amounts in them" (Porgera PR #6, 1963/64). Significant amounts of money were being made in Porgera; prior to June 30, 1960, around 2,500 Australian pounds was paid to the miners. After Robinson demarcated the new claims in the area in October 1960, the new miners mined 204 ounces of gold with a value of £2,500. By the end of 1960 about £7,000 had been paid out. More men than ever were now involved in the goldfields as well. McBride estimated that as of December 1960, of 420 able-bodied men in the area, 300 to 350 of them were working at alluvial mining, 250 to 300 were working on the indigenous claims, while Taylor had fifty men working on his claims (Laiagam PR #2, 1960/61). The fact that more people now had access to mining also led McBride to claim:

> Since the majority of the people have had access to gold work and since we have commenced to build the airstrip the peoples' attitude towards the Administration has changed considerably. The people have become co-operative and have now volunteered to work a week of each month on the airstrip. They just were not interested before. Income from gold and the need to get supplies into the spot have of course had much to do with this change in attitude. However, the place would probably "explode" if we withdrew direct supervision now. They are a greedy people and gold will make them more greedy.

The diversity of comments about the Ipili and wealth by patrol officers foreshadows some of the attitudes about wealth that are discussed at the beginning of chapter 7. For a people who, having the wealth of the forest at their doorsteps, are not reliant on a cash economy, attitudes toward money are fundamentally different than would be the case in capitalist-based societies. Moreover the shock of colonial authorities at "uncontrolled natives with bank accounts" demonstrates the facility with which Ipili miners were able to adapt to novel economic circumstances while simultaneously disrupting colonial stereotypes about recently contacted people. This same facility at negotiating colonial structures and institutions was also employed outside the sphere of mining.

Leadership, Colonial Punishment, and Missionization

While some Porgeran men were able to channel their newfound mining wealth into avenues of power, others unwittingly (at least initially) found alternative means to mining, such as being arrested and sent to jail in Wabag, to gain power and prestige. One of these men was Ben Penale's grandfather, Bope. In February 1999 I took a week-long break from fieldwork in Porgera to go to the National Archives in Port Moresby. While reading through a patrol report (Laiagam PR #3, 1954/55) there, I came across a mention of Bope from Tipinini having shot a man with an arrow right before the patrol entered Porgera and the fact that Bope had been jailed in Wabag. When I returned to Porgera, I asked Pes about the events that led to his father being arrested. He launched into a very long, complicated story with multiple characters from several clans. In essence, Bope's brother, Tapien, had been wrongfully accused of stealing food from a garden during a famine and was shot with an arrow by a clan brother of the owner of the garden. Tapien died a few days later from the wound. When an initial compensation (tapua) of eight half-sides of pig was given for Tapien's death (the larger compensation payment follows the tapua), Pes recalled, "Two men from Alipis [the alluvial mining area] came to claim the tapua. They both claimed to represent Tapien's father's and mother's clans. They were wearing trousers and we were all intimidated because of their trousers, and we didn't speak up to try and claim any of the tapua for ourselves." One of these men was Ipape, who had left with the Hagen-Sepik patrol in 1939.

A fight broke out between the Tokoyela (Tapien's clan) and the Pilisame (the garden's owner's clan) over the compensation, during which Bope wounded Lo (Ro in the report) with an arrow. He was arrested and sent to jail in Wabag for seven months. While incarcerated, Bope learned some Tok Pisin and saw cattle for the first time in his life. Pes noted, "Bope came back and told us that he had been forced to plant casuarina

Figure 2.3. Peter Enopi overseeing his daughter's bridewealth exchange.

trees the whole time around the government station. He had to follow around a 'big pig with two teeth sticking out of the top of its head that only eats grass and shits a lot' and gather its manure to plant with the tree saplings." When Bope arrived in Porgera, "he was carrying a huge sack full of the lids of food tins" from the patrol officers' meals. The shiny lids were so sought after as items of personal adornment that he was able to acquire a vast wealth in pigs and shells. "People wanted the lids so badly that they fought over them," Pes remembered. "The first thing Bope bought after this was a pair of trousers." With his new trousers and ability to speak Tok Pisin, Bope went on to become one of the key nonmining leaders in the Porgera valley.

Peter Enopi, who was killed in 2003 in tribal fighting, was another of the nonmining leaders in Porgera from 1963 until his death (figure 2.3). He had been sent to jail in Laiagam in 1957 for killing a man, was moved to Wabag to await his trial, and was incarcerated in Wewak until 1963. While there he became fluent in Tok Pisin, attended Lutheran church services, and gained knowledge about the colonial government's policies and practices. Upon his return to Porgera, he was hired as an interpreter and actively promoted the building of two Lutheran churches in the eastern Porgera valley. After Independence he served as the village councilor in Tipinini. He told me, "One of my daughters was born at this time, so I named her Mana [Ipili for 'law,' 'teaching,' 'custom'] because I was trying to teach all the people about the government."

Being in the church was another avenue for some men to gain prominence in the nonmining sector. Tisapa Yawe provided the following narrative:

One day word came to us that there were some white men making a ples balus [airstrip] at Laiagam [this would have been in 1958]. None of us knew what this thing—ples balus—was, nor had I ever seen a white man, so I decided to go see what was happening. Pasia, Peter Enopi's "father" [father's brother], had already left, so I followed him.

On the way to Laiagam, I met a Laiagam man coming to Porgera. His name was Pewalo and he said he was carrying misini pi ["mission talk" in Ipili] to give to people. I asked what misini talk was. He told me that if I went to hear it, I would no longer have a stone axe and wear leaf buttock covers, but I'd wear a nice shirt and trousers and carry a steel axe. He told me that misini pi was the church of the white men and if I got baptized [ipa mia, "take water" in Ipili] that I would go to live in the sky above the mountains. He said I would always be clean and have soap to wash my skin.

I told Pewalo to go on to Porgera and I'd help him with the misini talk when I got back from Laiagam. At Mulitaka I ran into Pasia, who was on his way back with another missionary from Laiagam. He and Pasia were going to bring misini pi to Pasia's clan at Yulindaka. I gave up the idea of going to Laiagam and returned to Porgera with Pasia.

I helped Pewalo build a church at Kiya when I returned. He gave sermons and we became Apostolic. In 1964 a white man came and made me one of the leaders of the Apostolic church here. He told the leaders to shave, gave us soap, and bought us trousers and a shirt to wear. The three of us [church leaders] were given metal cards that said we were mission men. We had these cards so that we didn't have to work on government work projects and we couldn't be arrested for not joining work parties. (field notes, April 1999)

Pasia had also encountered an Apostolic missionary and had built a church in Yulindaka. When Peter Enopi returned from prison as a Lutheran he convinced Pasia and the rest of his clan to become Lutherans. The Lutheran mission at Yuyan provided money for concrete and corrugated tin, and their new church was one of the first nontraditional material structures in the eastern Porgera valley.

Getting arrested and jailed or adopting Christianity provided means to become acculturated in the new colonial regime that men then used as a form of upward mobility. These processes also allowed for the alchemical transformations of the body that Porgerans were keen to initiate. The fact that Tisapa was told he would live in the sky above the mountains resonated with several myths in which the protagonist, after a series of trials, is whisked up into the sky world to live with the immortal sky people. *Tawe toko* (sky bridge) and *ati kenga* (mountains above) were terms used to depict this sky world. Early missionaries quickly capitalized on this indigenous sacred geography by using these terms to refer to heaven to hasten the process of proselytization.

to induce someone to convert one's faith.

Opening the Valley to Outsiders

De-restriction of the Porgera valley—the change in administrative status from uncontrolled to controlled—on July 1, 1962, brought a host of changes to the area. A permanent patrol officer and patrol post were established, as was an airstrip that could bring goods in for the newly wealthy Porgerans to purchase. Missionaries, who had been secretly sending in native lay preachers to proselytize (Laiagam PR #2, 1958/59), could now actively build churches in Porgera. Catholic, Lutheran, and Apostolic missions were built and run by white missionaries. Seventh Day Adventist missionization was run from Laiagam. By 1964, though, wealth from indigenous mining was becoming increasingly concentrated in the hands of a few men. A patrol officer reported that mining money falling "into the hands of a few . . . [results] in their material wealth and consequently power being far in excess of the majority who are still confined to traditional modes of existence" (Porgera PR #6, 1963/64). This patrol officer went on to note that despite the benefits of mining realized by a "fortunate few," life for most people was unchanged, "except that sporadic killings have now been curtailed." Another patrol officer noted that there were about one hundred miners who were earning good money, while the rest of the population was not (Porgera PR #1, 1969/70).

In the late 1960s, with mining wealth accruing among the elite, development benefits were still lacking. Of approximately one thousand children in the valley, only fifty were attending schools run by the missions (Porgera PR #4 1967/68). The government administration built a school at the end of 1968 that hosted standard grades 1 and 2. "The Porgeras [sic] are sensitively aware of the fact that Laiagam, Kandep, and Tari are developing more impressively than they, following the introduction of Councils to each of these areas, and also that the Porgeras are generally considered to be bushmen by other centres, a phrase which is not in their opinion complementary" (Porgera PR #1, 1969/70).

Besides bringing missionaries into the region, de-restriction also brought mining companies. Rio Tinto, Anaconda, and Bulolo Gold Dredging conducted exploratory work in the upper Porgera in the late 1960s. Porgerans were aware early on of the problems that could arise with large-scale mining development. During a 1969 patrol the patrol officer wrote, "Lectures were given to the people at all centres regarding mining, both local and company, along with the advantages of mining companies being involved in the area and the economic development that is possible under those circumstances. These talks were well, albeit somewhat mundanely, received" (Porgera PR #1, 1969/70). This officer further noted that the diamond drilling and geological explorations on Mt. Watukati made the Ipili miners fear "that this would result in the eventual disappearance of the gold bearing mountain, Waruwari [Watukati], and that gold would cease to be carried down the rivers—thus depriving them of their economy and living." He dispelled their fears, but he himself worried that future developments would indeed bring the issue to the surface again. As will be seen, Ipili miners were rightly worried about the mining company's impact on their gold livelihoods.

Papua New Guinea's Independence in 1975 did little to change life in Porgera, as throughout the 1970s and 1980s good money was being made by a number of people in the goldfields. Many Porgerans controlled their own alluvial leases and allowed kin and friends to work gold for themselves, sharing some of the proceeds with the lease owner. Mt. Isa Mines (MIM) acquired Taylor's claims in 1972 and started a sluicing operation. Porgeran miners worked on a tributary basis for MIM and received 75 percent of the proceeds (Handley and Henry 1987: 11). In 1975 Placer (PNG) Limited, a wholly owned subsidiary of Placer Dome, bought a two-thirds interest share in MIM's alluvial mining operation and also took over the sluicing operation. Most of the wealth earned by Porgeran miners at this time was going toward conspicuous consumption. Philip Gibbs (1977: 22), a Catholic priest-anthropologist, reported that in 1974, from wages and occupation fees paid by MIM, "at least eight plane loads of store goods arrive, including two loads of beer" each week, representing "about 5000 pounds of foodstuffs and 50 cartons of beer."

The Development of the Porgera Mine

By the late 1970s the development of a large-scale mine seemed promising,[6] so in 1979 Placer, MIM, New Guinea Goldfields, Ltd., and the Papua New Guinea state started a joint venture agreement, named Porgera Joint Venture (PJV), with each mining company holding a 30 percent interest in the joint venture while the state held 10 percent. Placer took over sluicing operations from MIM and started to drill adits in earnest to explore the potential gold resources. In 1982 the discovery of an ultra-high-grade ore zone in the heart of Mt. Watukati meant that mine development was certain to occur. Mineralogically, though, industrial gold-processing techniques were not sophisticated enough to separate Porgera's gold from the iron sulfide matrix in which it occurred (Jackson and Banks 2002: 73–81), and it would take a few more years before a processing method would allow Porgera's gold development to begin.

Before any actual development could occur for the building of the Porgera mine, PJV needed to complete a number of impact studies on both the people and the land as part of the development plan (see Filer 1999). National requirements for development under PNG's Mining Act of 1979 stipulate that the owners of land impacted by mining within a Special Mining Lease (SML) area are to receive compensation for damage to crops and permanent improvements on the land, annual occupation fees, and royalties worth 5 percent of 1.25 percent of the value of mine production. Local people in the area, including landowners, are to receive compensation for any damages outside the SML area and preferential employment and business development contracts. Provincial governments receive the other 95 percent of the 1.25 percent of royalties. The national government receives a variety of taxes as well as the opportunity to purchase a portion of the equity in the mine (Jackson and Banks 2002). For local people, the greatest benefits come from having land located within the boundaries of the SML. In the development discourse of PNG, these people are considered the "landowners" associated with any development project. The SML at Porgera was 2,227 hectares; genealogy studies indicated that approximately 6,763 people (Finlayson 2002: 41) from seven clans could claim some sort of traditional rights within this area. Due to the cognatic nature of Ipili society, many of these 6,763 were duplicates, even triplicates or more, as individuals can affiliate with multiple clans. The seven clans comprised twenty-three subclans, and the landowners were to assign a representative from each subclan to negotiate the Porgera development package. Twenty-two men and one woman were appointed to serve on the Landowner Negotiating Committee, which would deal with the mine and the state. By May 1989 the Landowner Negotiating Committee had signed a set of agreements called the Porgera Agreements (Derkley 1989), which detailed a compensation package that the landowners would receive for the loss of their lands from mining.

Stemming from the Bougainville crisis (May and Spriggs 1990), the Porgera mine was planned in an era when landowner concerns were a central part of the development process. The final set of agreements

Table 2.1 Royalty distributions (in %) for the Porgera Mine

	1990–1995	1995–present
Enga Provincial Government	77	50
SML Landowners	8	15
Children's Trust Fund	10	10
Porgera Development Authority	5	5
Young Adults Fund	—	8
Porgera Landowners Association	—	12

Royalty shares increased from 1.25 to 2 percent of total value of mine production in 1996.

went beyond what was just part of the national requirements (see table 2.1). In addition to compensation and annual occupation fees, Porgera landowners were to receive 23 percent of 1.25 percent of royalties of the value of mine production (the other 77 percent going to the provincial government; in 1995 this changed to a 50/50 split, and in 1996 the royalty share increased to 2 percent). Of the 23 percent, 8 percent was to go to the SML landowners, 5 percent to the Porgera Development Authority, which would oversee development projects in the valley, and 10 percent went into a trust fund for the children of the landowners. In 1995 the distribution of royalties increased to 15 percent for the SML landowners and for the Porgera Development Authority, and the children's fund remained the same, but two new entities were created: the Young Adult's Fund (8 percent) and the Porgera Landowners Association (12 percent). The Young Adults Fund was a direct result of the Bougainville case, in which younger people, who had been left out of royalty payments, were as adults rightly upset over the social and environmental impacts of mining (Filer 1990). The Porgera Landowners Association represents the twenty-three SML subclans. According to a consultant hired to study the benefit streams from the mine, "There is no record of the objectives of the association or how they intend to spend the money" that they have received, which up to 2009 totaled K30.6 million (Johnson 2012: 46–47). A special support grant totaling 1 percent of the value of mine production was paid to the provincial government for giving up their share of royalties (from 95 to 77 percent) to the landowners. This amount ended up being six to seven times more than the royalty shares would have been (Jackson and Banks 2002: 167) and went toward various infrastructure improvement projects. The landowners and the province could buy up to 10 percent equity in the mine from the national government's share in the future (they now each hold 2.5 percent). Preferential employment at the mine would go to landowners, other locals in the mining area, other people from the province, and Papua New Guineans, in that order.

The increases in benefits over national standards negotiated by the Landowner Negotiating Committee highlights the shrewdness that Porgerans had learned in their years of dealing with expatriate miners, transnational mining companies, and national elites. At the same time that the landowners strengthened their position against the provincial and national governments, conditions were also set to create greater disparities in wealth between the landowners and the nonlandowner Porgerans, the bulk of the population in the valley. The Porgera Agreements were officially signed by the national government, the Enga provincial government, and the Porgera landowners. While PJV was not a signatory to the agreements, they were considered a "silent partner" for the Special Mining Lease at Porgera would not have been granted had PJV not acquiesced in the demands set forth in the agreements (Derkley 1989). In addition to the generous compensation package for the landowners, the other major parts of the Agreements centered on bringing services (i.e., air transportation, hospitals, schools) to the area and to the development of a modern town, Paiam. On the cover of the Agreements is a cartoon showing an Ipili man dressed in traditional attire, with a leaf buttocks cover, a traditional wig headdress, bird of paradise feather decorations, and a mother-of-pearl shell around his neck (figure 2.4). Considering that by 1989 there were few if any Ipili who still routinely dressed like this, this caricature highlights what most people outside of the Porgera valley felt about the Ipili. Most Engans expressed bewilderment to me about how such a "backwards, primitive" people could become the beneficiaries of a world-class gold mine. In the cartoon, floating just out of the reach of the Ipili man, are two pieces of paper. Although he is smiling, it appears that he is chasing the two pieces of paper floating above him. For many people this cartoon may be more apt than it first appeared, as they complain that they are still chasing after the promise of development that was offered to them by PJV and the government.

The mine officially opened in September 1990 and by 1992 was the third richest gold mine in the world (Biersack 2006a). By this time, though, the effects of riverine tailings disposal were starting to impact the "second gardens" of alluvial gold downstream from the mine. As tailings and waste rock began to cover the lucrative alluvial beds, formerly rich and powerful landowners in the lower Porgera, such as Puluku, saw their wealth being transferred to the upper Porgera SML landowners. The

Figure 2.4. Cartoon of Ipili man on the cover of the Porgera Agreements (Derkley 1989).

lower Porgerans started to organize under the leadership of Opis Papo, an eastern Porgeran who obtained a university degree in geology with the assistance of PJV and was employed at the mine, and formed the Porgera River Alluvial Miners Association, which in 1996 successfully negotiated a compensation plan for the loss of the alluvial beds in which quarterly payments are made to key landowners based on the actual tonnage of tailings that are released into the Porgera River. In 1997 Opis Papo was elected a member of Parliament from the Porgera-Lagaip district on the campaign platform of being "a sworn enemy of the mine" and ensuring that mining development would benefit all Porgerans and not just the SML landowners (Opis Papo, personal communication, December 8, 2006).

By the time I arrived in Porgera in 1998, factions of enfranchised SML landowners and disenfranchised non-SML landowners were beginning to form. The resource frontier that initially seemed so promising to many in Porgera was starting to resemble frontiers the world over, with great wealth for a few and social and environmental disruption for the rest.

Part II

Indigenous Philosophies of Nature, Culture, and Place

The following three chapters describe indigenous understandings of ecology and human-environmental interactions. Anthropologists have long been concerned with how to write about nature and culture in other societies when, as Marilyn Strathern (1980: 219) writes about the Hagen people of highlands Papua New Guinea, "There is no culture, in the sense of the cumulative works of man, and no nature to be tamed and made productive." The task for environmental anthropologists, then, is to capture the complex interplay of cultural actions and attitudes that are in constant dialogue with geological, climatic, and ecological processes (see Escobar 2008: 42–43). The natural world loosely constrains the cultural meanings that humans give to it without actually determining the cultural formations that can arise in given environments. To speak of a nature outside of culture, or a culture outside of nature, is problematic as they are mutually constitutive (Descola and Pálsson 1996; Escobar 1999; Ingold 2000). Our central concern, then, is seeking to understand the processes by which the environment is constituted by human activities and what impacts environmental changes have on those activities—an approach that is encapsulated by the term *landscape* in the social sciences (Bender 1993; Crumley 1994; Hirsch and O'Hanlon 1995).

Studying how material and discursive cultural practices constitute the "natural" world and vice versa, geographers have long utilized the concept of landscape to capture the fact that culture and nature are inseparable entities (Bender 1993; Braun 2002; Massey 1994; Smith 1996). Anthropological interest in the relationship between landscape and culture began in the 1990s (Basso 1996a; Rodman 1992). At first rooted in more symbolic and semiotic approaches to human-environmental relations, landscape and cultural analyses were eventually influenced by critical studies in geography (Blaikie and Brookfield 1987b; Bryant 1992; Peet and Watts 1993), what today has become the field of political ecology. From an early interest in landscape, environmental anthropologists have shifted to examining the relationships between place and ontology. As I mentioned earlier, Porgeran ontological systems are characterized by an interconnectedness of the total environment (e.g., geology, ecological processes, nonhuman others) and human affairs. In part II, however, I intend to disentangle some of this interconnectedness by exploring critical and foundational concepts in a Porgeran philosophy of ecology. I center my analysis around three principal concepts that Porgerans discuss when talking about human-environmental relations: land, people, and spirits. Chapter 3, "Land: *Yu*," provides an analysis of conceptual divisions that Porgerans make about the landscape in the cosmological sense of how the world is depicted, as well as how the landscape is divided along a tripartite division based on elevation, with attendant practices of resource utilization, social rights, and relations with various nature spirits. Land is also the basis for forming social groups, and in chapter 4, "People: *Wandakali*," I examine the relationships between production (*piape*, "work") and kinship—that is, how social formations are shaped by land use practices and how kinship networks facilitate particular land use patterns. Chapter 5, "Spirits: *Yama*," provides the final element for understanding socio-ecological processes in Porgera. Spirits are active participants in both natural and social systems and provide the means to access knowledge about the intentions of people and natural entities in the landscape (see also Jacka 2010).

My intention in separating these concepts—land, people, and spirits— is to examine in detail a Porgeran philosophy of ecology in order to understand change in a coupled human-ecological system. It is critical to emphasize that the conceptualizations I present here are not "traditional" in the sense that people no longer think this way and have replaced these ideas with "modern" ones. Nor are these concepts timeless wisdom handed down by the ancestors. They are an ever-changing, flexible, and adaptive set of ideas that shape how Porgerans engage with their environment and with human and nonhuman actors in their world (cf. Berkes 2008). As will be seen, the powers that some of these nonhuman entities have is undergoing debate in an era of Christianity, but by and large the following three chapters detail ideas common among most Porgerans about how the world works. Most important, unlike Western concepts of environmental ethics that rely on notions of respect, reverence, awe, stewardship, and so forth, Porgerans do not revere or respect nature. What a Westerner would call "nature" in Porgera has been shaped by a long-term reciprocal engagement between people, the land, and spirits. The central issue is understanding the reciprocity between people and the world around them.

Knowledge of place and landscape is a critical component of social-ecological resilience. Porgeran social groups and land-tenure dynamics are inherently loosely structured and flexible. People are expected to utilize diverse environments and the lands of other groups to enhance their livelihood opportunities. Knowledge of kinship networks and the myths associated with certain areas provides a person with the possibility to access lands held by other groups. Nevertheless spirits—nature spirits, ancestral spirits, and the spirits of slain enemies—constrain the ability of a person to fully utilize all potential environments and kinship linkages.[1]

CHAPTER 3

Land: *Yu*

Geology and Human Affairs

Late one night in November 2006 I awoke to a rumbling, crashing sound that seemed to emanate from the very house I was sleeping in. Initially startled, I quickly realized that it was a landslide roaring down the face of Mt. Pauwite, a kilometer or so behind my house. I soon fell back asleep as the persistent rain hammered on the tin roof, almost drowning out the occasional boulder that could still be heard tumbling down the sheer face of the mountain. The next morning I stood outside assessing the damage that had been caused by the landslide. Just two days earlier my research assistants and I had been at the base of Mt. Pauwite cataloguing tree species in the remnants of second-growth forest that were quickly being cut down to make gardens. I knew that there were gardens right at the base of the mountain, and I was curious to see if any damage had come to people's crops and the fruits of their hard work.

Solomon Kaipas, one of my research assistants and the son of the Lutheran pastor whose house I was sleeping in, joined me, and we stood looking at the new brown gash in the white limestone of the cliff face caused by the landslide. We weren't surprised to see that the landslide had indeed slid right into the uphill portion of someone's garden. That gardens are even being built at the base of Mt. Pauwite is a noteworthy sign of the recent changes in subsistence practices in Porgera. The southern edge of the Porgera valley is hemmed in by high limestone cliffs, up to 1,000 meters above the valley floor in some places. The slope of the land at the base of these cliffs is around 40 degrees. It was hard to believe that the heavy rains hadn't already washed the sweet potato mounds right off the slope. Moreover at 2,400 meters above sea level, the gardens at the base of Mt. Pauwite were nearly at their altitudinal limit for producing crops. Nevertheless people continue to move up into the high-altitude rain forest and plant their gardens in such precarious locations to be near the highway that services the Porgera Mine, despite the fact that firewood and building materials are becoming more scarce and gardens take longer to produce vegetables.

Regarding the landslide, Solomon said, "This isn't good. It means someone's going to die soon." He explained that Mt. Pauwite is the mountain of the Maiyuni clan. Whenever a landslide occurs on Pauwite, it means that a Maiyuni man will die. He pointed out two other recent landslides that also scarred the cliff face. The one to the right of the previous night's landslide occurred some months ago, and within a week a Maiyuni man was killed in a tribal fight. Two months earlier the landslide to the left of the most recent one had occurred, and yet another Maiyuni man had been killed in fighting. This most recent landslide meant it was someone else's time to die. I didn't doubt Solomon. Death is an ever-present reality for Porgerans, and given the dense and effusive network of social relations that people share, it would hardly be surprising if the next person to die in the region could trace some sort of connection to the Maiyuni.

Prior to the late 1980s no one had lived anywhere near the base of the limestone cliffs that demarcate Porgera's southern boundary. Not only is the soil constantly waterlogged and of poorer quality than land lower in elevation, but there are numerous spirits that inhabit the high rain forest that potentially threaten human lives. Decisions about where to live, garden, and hunt are shaped by concerns about the various kinds of spirits that occupy particular landscapes. Spiritual entities also live within the numerous mountains, and their movements cause the frequent landslides and earthquakes of this tectonically active region. But with Christianization and the building of a road in the upper rain forest, the landscape has become less enchanted with spirits and more attractive for human settlement. Notwithstanding this disenchantment, spirits in the landscape are still very much a part of contemporary life in the Porgera valley, as witnessed by people's belief that landslides are portents of impending doom. Two other landslides that occurred in 1999 while I was in Porgera, one on Mt. Paiam, just to the west of Mt.

Pauwite, the other to the east, at the base of Mt. Pauwite, caused serious consternation in the affected communities due to fears of displeased spirits that would bring further harm.

To talk about land in Porgera is to talk about kinship. This chapter focuses on the material and ideological aspects of the land. The following chapter, while ostensibly about people and kinship, is also very much about land, as the two are deeply intertwined. The relationship between kinship and the land is captured by a word in Ipili, *tane*, which has four meanings: self, belonging to a place, own, and relatives; *nambana tane* means "my self," "my place," "my own," "my relatives." To be from a place is to be of that place and of the people whose labor shaped the land. To say that you are from somewhere is as much a statement about sociality as it is about geography. There are socially determined (i.e., open to debate and contestation) boundaries for each clan in the upper and lower rain forests and in lands between the high-altitude and low-altitude forests. To "be" a clan member is to have use rights in the areas where one's clan's lands are located—to be an "owner" of a place (cf. Stasch 2009). As will be seen over the next few chapters, the goal in Porgeran society is to maximize the number of clans that one is a member of in order to gain access to a variety of lands throughout the valley.

Indigenous Conceptions of the Landscape

Ideas about the land, or *yu* in Ipili, have been shaped by the interplay of heavy precipitation and geological processes. In the Porgeran worldview the land is the middle layer of a three-part universe, sandwiched between a sky world and an underworld. A foundational concept in Porgeran cosmology is an anthropomorphized earth that ages, sustains life, and dies, much as humans do (Biersack 1987, 1991, 2005, 2011b). Tisapa Yawe told me a story about a woman who was the mother of all Porgerans and who discovered agriculture. In this story Tisapa noted that when she first started farming "the world was very young so all the crops grew very well, just like a young woman who produces good children; old women, on the other hand, don't give birth to good children." Porgerans have two means to ensure that the earth remains fertile "like a young woman." At a practical, everyday level, shifting cultivation is one means for countering the degradation of the land. The other way is through rituals that propitiate spirits associated with the land (see chapter 5), although these are in decline and require large-scale coordination across clans.

Porgeran concepts of the land partially emulate geological under-standings of plate tectonics. In Porgeran cosmology there are a series of "earth joints" (*yu kimbuni*) that tie the land together. Spiritual beings hold together the earth joints and prevent the world from breaking apart. To demonstrate the concept to me, one man placed four small sticks in the palm of his left hand with their ends radiating out in the cardinal directions. With his right index finger he held all the sticks in place in his palm. "If the spirits let go of the earth joints," he said while removing his finger, causing the sticks to fall to the ground, "the ground ends" (*yu koyo peya*). Ritual propitiations are made to the spirits that hold the earth joints together whenever declines in fertility are seen in the land or people or during severe climatic events, such as droughts.

After fourteen generations of people have been born, ritual propiti-ations are said to no longer be effective, and the earth is destroyed and a new one arises. As Muyu Yakati described it:

> There is a time when the ground will end. At this time Helape and Kelape [two ancestral spirits that keep the rivers flowing by removing trash from them] will remove the *lokai* wood [*Dodonaea viscosa*] from the primordial fire, and all of the fires in the world will go out. The brothers will kill the pig that they have been sitting on, and they will let go of the earth joints [*yu kimbuni*]. Carrying pig meat they will come to meet the man from Tuandaka who will let go of his earth joint [see chapter 5]. Together they will all meet at Aiyapo Puka [a swamp on the backside of Mt. Asienda]. At this time, all of the people who are cutting up pig meat, their [bamboo] knives will break. There, the three will eat pig, dance, and beat their *kundu* [hourglass-shaped drums, figures 3.1 and 3.2], and the world will fall apart and we will all die. (field notes, January 1999)

I heard over a dozen variations of Helape and Kelape's role in ending the world by letting go of the earth joints. Muyu himself told me the story four times between January and November 1999. After the fourth telling, Peter, Muyu's son, commented, "I think that this [story] is just a *tok piksa* [parable]. Our ancestors didn't know about the outside world. They were talking about people from the farthest parts of the highlands they knew of coming to Porgera to eat pig here, which would cause the end of the world. But they were really talking about people from all over the world eating pig here. They weren't talking about a real pig; they were talking about the gold mine and people eating its wealth."

Not only does the land have joints; it also has a skin and bones. When Porgerans talk about landownership they refer to this notion as a

Figures 3.1 and 3.2. Men carving hourglass-shaped *kundu* drums.

particular subclan "holding" the yu *kulini* (ground bone) of the land that is associated with that clan. Individual clan members, however, "hold" and occupy the yu *umbuaini* (ground skin) above the ground bone. Topsoil, trees, and plants compose the ground's skin over which individuals have usufruct rights. As individuals are transitory and mortal, opposed to the ideal of the permanent clan, there are no actual concepts of landownership. In this sense land cannot be sold since the people merely use it. The clan, which is composed of living people, their ancestral spirits, and their unborn descendants, cannot sell the land as it is necessary for the production and reproduction of the group.[1] With skin, bones, and joints, the land is an anthropomorphized entity. Discussing rituals to revitalize the land, Yakob Kuraia, a former ritual expert, said in reference to why the ground would soon end (in late 1999), "The land no longer produces like a young woman. She is old now, and tired."

Previously I documented (Jacka 2001) a ritual that Porgerans claimed was no longer performed. The ritual was conducted in 1999, shortly after agreements had been completed about housing mine workers in the valley rather than flying them home on their breaks. The agreement stipulated that the houses would be constructed in the new town of Paiam. The town is named for the clan whose land, depending on one's perspective, was given, sold, or loaned to the government. During the construction of the houses, mysterious events started to occur in the Paiam Supermarket. Some mornings workers would come into the store, and there would be empty but unopened cans of Coca-Cola and tinned meat and fish on the shelves. Since the supermarket was built on the site of a sacred pool where the spirits of deceased ancestors dwelled, the Paiam clan interpreted the empty cans as evidence of the displeasure of their ancestors. To placate them they held a kolo ritual in which pigs were killed in the supermarket parking lot and their blood was allowed to drain into the ground to feed their ancestors. During the ceremony a large chunk of limestone broke off the face of Mt. Paiam, causing a large rockslide visible to all the participants. Later, people claimed that the ancestors were upset about the land transfer from the clan to the government for development purposes. The interpretation reflects people's consternation over the transfer of land. The quasi-government agency that oversees the development of Paiam, the Paiam Management Company, argued that the monetary compensation that the Paiam received for the land was a payment. Many of the Paiam, on the other hand, believed that the payment was not a one-off affair indicating a sale but was a payment for use and occupation. How, they wondered, could a landscape alive with ancestral spirits be alienated forever from the clan?

Opposed to a living, mortal earth is an eternal sky world peopled by immortal sky people (tawe wandakali) and the deity-like figures of the male sun (nai) and female moon (ana). Nai, the sun, is an omniscient figure that guides people's moral behaviors and has the ability to punish people for social transgressions. The sun "sees and knows everything" (Biersack 1987: 188; cf. Gibbs 1977: 9–10) and is responsible for giving humans mana (laws, custom). The fontanel of the infant is the spot where the sun transmits mana to the child at conception. Ipili speakers refer to the omniscience of the sun by the phrase nai lene andeya (the sun's eye sees/knows). In Ipili the verb "to see" (anda) is also the verb "to know," and the sun's eye is able to see/know "inside caves, the forest, inside houses, the good and the bad, women and men . . . everything," as one Paiela elder reported (Biersack 2011b: 231).

Ana, the female moon, has the power to control human and vegetative growth. One afternoon I was standing in a garden that one of my friends, Julie Muyaka, had just planted with sweet potato cuttings. The newly planted vines lay wilted on top of the sweet potato mounds. Julie pointed to the wilted vines and said that the sun was too powerful, it killed things; however, when the moon "looked" at the cuttings they would start to grow. Biersack (1987: 186–87) reports that the Paiela valley Ipili say that "animals, humans, even sweet potato tubers, grow as they 'sleep.' . . . All growth magic is performed during the colder and darker parts of the day, at night and especially in the morning . . . 'lest the sun see.'" Another notion of growth needing to take place in the dark comes from rituals that men use to grow their hair. One morning in 2006, in a forest adjacent to his garden, I ran into a man in his mid-forties, Simon Solei. Simon had stopped in the trail and was adjusting a thin wig of moss over his hair. I asked what he was doing with the moss and he explained that he had just "put" some growth magic on his hair that morning as he had recently begun to worry about going bald. He needed the moss to protect his bespelled hair from the sun; otherwise the growth magic wouldn't work. I often saw men using all manner of items—bark, plastic bags, moss—to protect their hair from the sun and make their hair growth magic more efficacious (figure 3.3).

The sun and the moon are transient figures. After traversing the sky they sink into a limestone cave far to the west and travel underground to fly out of a cave far in the east and begin their journey across the sky again. While underground, they are in the realm where the spirits of deceased humans live (*talepa anda*, "spirit house"). The underground world is riddled with caves and tunnels through which the sun, moon, and spirits move.

Cosmologically, then, the universe is divided into three parts: an immortal and eternal sky, a mortal and relatively impermanent land surface, and an underground world that is unknown to humans but populated by mostly immortal spirits. This tripartite division is also conceptualized in the cognitive perception of spatiality in Porgera. In Ipili there is a system of relative spatiality that defines whether someone or some place is higher in elevation, at the same elevation, or lower in elevation than the speaker. This perception of relative high, same, or low elevation is combined with a perception of relative distance, determined as near or far. To talk about spatiality or location, then, requires a speaker to place herself or himself in three-dimensional space. Schematically it looks like this:

lipi: nearby at a higher elevation
iyu: far away at a higher elevation
ambi: nearby at the same elevation
amo: far away at the same elevation
napi: nearby at a lower elevation
nena: far away at a lower elevation

Regionally the system of spatiality overlays the major geomorphic features of the area. The high-altitude karst lands above Porgera are *iyu*;

Figure 3.3. A man covers his hair to protect it from the sun.

hamlets farther upstream along the Lagaip River are iyu. Iyu in fact encompasses the entire upper watershed of the Lagaip River. The nearby features extend to within a kilometer or two of the speaker, or a location a person could reasonably walk to in a few hours. The conceptualization of features at the same elevation that are far away (amo) extend only to locations within the Porgera valley. So even places in other provinces that are fairly close in elevation to Porgera are not amo. Once beyond the headwaters of the Lagaip and the boundaries of the Porgera valley, every other place is considered nena—far away at a lower elevation—even if in absolute terms the places are higher than or of a similar elevation to Porgera. With the exception of heaven, which is iyu, everywhere else in the world is perceived as far away at a lower elevation—America, Australia, even Tibet.

The category of nena—places far away at a lower elevation—also encapsulates the notion that there is a spatial aspect of morality encoded in the land. There is, in essence, a "moral topography" (Ballard 1998; Dove and Kammen 1997) that Porgerans attribute to the landscape. Several scholars have stressed that among the Huli ritual power is perceived to come from peripheral groups and places adjacent to the Huli homeland (Clark 1993, 1995; Frankel 1986; Goldman 1983). Among the Ipili too the periphery is a place of great power and, potentially, danger. However, places higher in elevation are generally sources of growth and nurturance and are considered benign. Places lower in elevation are fecund and rich in resources, yet dangerous and generally feared. The threefold division of high, middle, and low is also expressed in terms of how the land's surface is conceptualized in more material terms. Porgerans divide their landscape into an upper rain forest (aiyandaka), a lower rain forest (wapi), and an area between the two where houses and gardens are located (andakama).

Yu Iyu: Land of the Sky People

The Porgera River begins in a high-altitude swampy grassland about 9 kilometers southwest of Porgera Station (map 3.1). In satellite images the grasslands look like long, narrow brown islands of parched earth surrounded by lush green rain forest. The grasslands (puka) comprise mostly sedges and other swamp plants. A few tree ferns dot the otherwise flat swamps. The grasslands are naturally occurring, as at 3,200 meters above sea level few people spend much time in this cold, wet terrain. High ridges covered by moss forest rise another 500 meters in elevation between the islands of grasslands. This is tower and cone karst country,

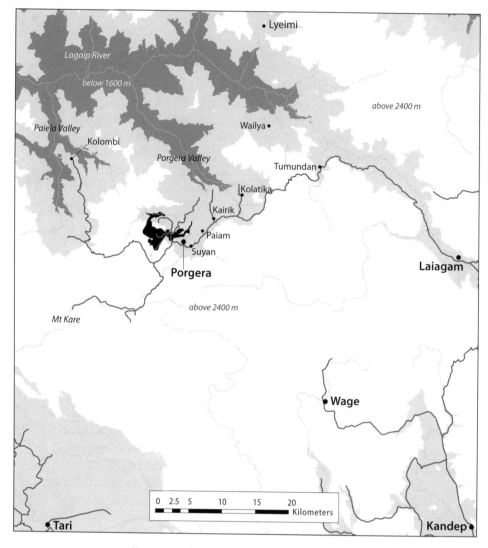

Map 3.1. Regional map of western Enga Province.

a tortured landscape of limestone sinkholes, ridges, caves, and tunnels (Löffler 1977: 22). Here rivers disappear, run underground for several kilometers, and reappear. Stories abound of hunters who come up here to find *kaima* (flying foxes; *Dobsonia magna*, great bare-backed fruit bat; Flannery 1995: 351–52) and are never seen again. In 2000 I met the son of a man named Anakali who fell into one of these sinkholes and was unable to climb out. He stayed alive for several months by shooting flying foxes, retrieving his arrow, and eating their meat raw. Eventually other hunters found him. They say he had become wild and was completely covered with hair—an unlikely prospect, but Anakali vowed never to go up to the high country again, and in fact moved his family into the lower reaches of the Paiela valley to remain far from the high karst country.

The grasslands and moss forests south of Porgera are part of a 3,000-square-kilometer limestone plateau that separates the Porgera and Paiela valleys from the Tari region, where the lands of Huli speakers are located. Traditionally this land was not occupied and was claimed by no one. A few trade routes cut through it that linked Porgera to Tari, allowing Huli to trade tree oil and pigs for packets of Ipili manufactured salt. In 1938 the part of the Hagen-Sepik patrol led by Jim Taylor nearly starved to death attempting to travel across this cold, wet plateau from Tari to Porgera. Taylor had mistakenly assumed that large numbers of people would have rich gardens of sweet potatoes to offer the explorers. Only after six tons of supplies were dropped by air was the danger of starvation overcome (Gammage 1998: 86), at which point the patrol turned back to Tari, failing to reach Porgera from the south. In the late 1980s the issue of who "owned" these lands became paramount when gold was discovered at Mt. Kare in the center of this plateau, 9 kilometers west-southwest of the Porgera headwaters (Ryan 1991; Vail 1995). Ipili, Huli, Duna, and Enga all ascended en masse to the goldfields of Mt. Kare. Teachers and health workers quit their jobs to participate in the gold rush, and whole areas became depopulated as people moved to the goldfields. By November 1988 approximately eight thousand people were mining for gold with picks, shovels, and their bare hands (Vail 1995: 345). It is estimated that K100 million to K150 million was taken out of the Kare area in the couple of years that the gold rush lasted.[2]

The high limestone plateau between Tari and Porgera, while an important trade route, was also a dangerous region full of spirits and other malicious beings. Tombeama (border spirits) guarded the high passes and would steal pigs and other trade goods from Ipili and Huli traders and in some cases would even kill people. The Mt. Kare region was also associated with a giant python spirit called Taiyundika who occasionally harmed travelers. On the track outside Porgera that led to Tari a large limestone sinkhole served as a ritual offering site to Taiyundika, into which people could throw gifts of meat and produce to ensure their safe passage. Jeffrey Clark (1993, 1995), Aletta Biersack (1995b, 1999b), and Holly Wardlow (2001) have all written extensively on the relationship of Taiyundika to Mt. Kare's gold. People say that the gold from Mt. Kare

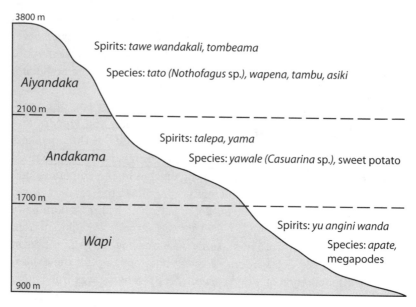

3800 m

Spirits: *tawe wandakali, tombeama*

Species: *tato (Nothofagus* sp.)*, wapena, tambu, asiki*

Aiyandaka

2100 m

Spirits: *talepa, yama*

Andakama

Species: *yawale (Casuarina* sp.)*,* sweet potato

1700 m

Spirits: *yu angini wanda*

Wapi

Species: *apate,*
megapodes

900 m

Figure 3.4. Schematic of Porgeran ecological zones.

comes from the feces or shed skin of Taiyundika. As Taiyundika travels around underground through the caves and tunnels in the limestone, he leaves behind gold deposits. Taiyundika is also linked to Huli ideas about the creation of the world (Clark 1995), and the removal of the gold thereby threatens the entire universe. The stories of Taiyundika and the struggles over Mt. Kare highlight important concerns about the ways that landscape is constituted in the Porgeran worldview and the consequences of the colonization of these landscapes by outside forces.

The limestone plateau and the slopes of the steep mountains that demarcate Porgera's southern boundary down to about 2,200 meters above sea level are called aiyandaka, which literally translates as "the place of beauty" (figure 3.4). In the recent past it was where the initiation rituals for young men were held and where earth renewal rituals were performed, and it is the rain forest zone associated with the tawe wandakali (sky people) of Porgeran legend. The elevation of 2,200 meters is an approximate boundary because it is not elevation alone that Porgerans use to categorize the altitude zones. Soil quality is another indicator, particularly as it pertains to the wetness of the soil. Women remark that when removing sweet potato tubers from mounds, aiyandaka soil is muddy and sticky and remains on the tubers. In the andakama (the middle elevation zone of houses and gardens), by contrast, the soil is more dry and friable and is easily brushed off the tuber. Various tree

Figure 3.5. Harvesting pandanus nuts in the upper rain forest.

species are other indicators that mark the aiyandaka zone. Two of the most common are *tambu* tree ferns (*Cyathea* sp.) and *wapena* wild nut pandanus trees (*Pandanus brosimos*). Another qualitative marker people use to define the upper rain forest zone is coldness. If caught in the aiyandaka in an afternoon rain, one gets cold and wet, whereas elsewhere one gets merely wet.

The sky people are the caretakers of the upper rain forest. The marsupials of this zone are called the "pigs" of the sky people. The sky people plant the wapena pandanus trees (said to be the "sweet potato" of the sky people) and the *asiki* giant taro plants (the "taro" of the sky people, inedible for humans; *Alocasia macrorrhiza*; *wel taro*, "wild taro" in Tok Pisin). If a human cuts down a wapena to get to the nut pod (as the moss-covered, slippery trunk cannot be climbed), that person is obligated to cry and mourn for the tree as if it were a human or else face the wrath of an angry sky person. Wapena are crucial during the height of the rainier season (November to March), as much of the Porgera area experiences a sweet potato famine due to the waterlogged soil that prevents good development of sweet potato tubers (figure 3.5). The sedge grass swamps that provide the materials for sky women's grass skirts are also planted by sky women, and Porgerans approach these swamps very carefully lest they come upon a sky woman gathering sedge grass for her skirt and startle her, making her angry.

Sky people are largely benevolent in that they allow people to hunt the marsupials of the high forest, gather the wapena nuts, and (in the past) harvest the sedge grass from their swamps. However, they can also be wrathful when earth-bound people transgress the moral precepts of moderation and proper behavior. Loud yelling and laughing, overhunting, wanton destruction of flora and fauna, and defecating on a trail were all mentioned as acts that can cause a sky person to become angry and prevent the marsupials and pandanus from being found by "hiding" them or cause the perpetrators to slip and break a leg, fall to their death, become lost, or have their hunting dogs run away. In December 2006, while cutting transect lines in the upper rain forest to count and catalogue tree species, Wanpis, one of my research assistants, cut down an asiki giant taro plant. His elder brother, Solomon, chastised him, pointing out that this was the taro of the sky women and now they would make it rain, as the sky people control the rain, thunder, and lightning. Not surprisingly the following day it poured all day long.

Besides providing freely of their pigs and sweet potatoes, sky people also figure quite prominently in myths as bestowers of great wealth for certain lucky individuals. In these myths a man gathers a wapena pandanus nut from the high forest, but upon opening it finds either a mother-of-pearl shell or a pig rope (or both) inside. The shell and the rope have been magically bespelled by the sky people such that the shell attracts other shells to itself—like a modern-day savings account, according to one elder. The pig rope magically multiplies one's pig herd such that every morning the rope's owner will awaken to discover an additional pig in his pig house. In these myths the men who possess such items soon become the wealthiest people in the region. Some men still perform a kind of wealth-accrual magic called *takia*. Takia is taught individually by elder men who are sought out and paid by a younger man who wants to become wealthy. Part of the ritual involves the bespelling of objects called *takame*, which, like the pig rope and pearl shell, work by drawing wealth to the possessor. One of the strongest and most important takame objects in eastern Porgera which a man allowed me to see was a single shell from a *tapuna* nut pandanus tree (*Pandanus julienetti*). Unlike most takame objects, but like famous wealth items from island Melanesia (Malinowski 1984), this takame was named and highly sought after by other men. Wapope Ambiambi, who currently owned the takame, related that it was the first takame ever used by humans in Porgera and was given to its original owner by a sky woman.

Yu Amo: *In the Zone of Houses and Gardens*

Literally the andakama means "the place of houses" (*andaka*) and "cleared spaces" (*ama*). In actuality there are house sites, cleared spaces that serve as meeting areas, churches, trade stores, new and old gardens, some remnants of primary rain forest, and much second-growth rain forest. The andakama lies between about 1,600 meters above sea level at its lower extent and up to about 2,200 meters at its upper extent. It sits in the optimal zone between the malaria and deadly spirits of the lower rain forest and the cold and frequent rain that inhibits crop development in the upper rain forest. Consequently it is the major site of production and reproduction in the social-environmental context of Porgera. The spatiality of these ecological zones also encodes a specific temporality. The high rain forest is associated with youth and vitality as that is where youthful beautification rites occurred. The middle zone of houses and gardens contains reproductively and productively active adults. It is where children are conceived and born, where gardens and houses are constructed, and where exchange activities take place. The lower rain forest is linked to old age, decay, and death as it is the resting place of human souls and is associated with socially destructive spirits.

The focus of social life in the andakama centers around the ama, the cleared spaces where public exchanges are held and, in their quotidian manifestation, where women sell secondhand clothes, cook small pieces of mutton and flour to sell, and sit and weave net bags and watch the daily activities of the hamlet (figures 3.6 and 3.7). Men primarily use the ama on a daily basis for playing cards and sitting in small groups watching people go about their business. Mornings and afternoons are busy times in the ama; during midday they may be completely deserted except for small children playing games. Most afternoons, when it is not raining, people bring court cases for the big men and village court magistrates to decide on. These are often small-scale affairs involving family members failing to share money with one another or someone's garden being damaged by a pig.

Despite the fact that there are "village" councilors and "village" court magistrates in Porgera, people actually live in hamlets. Close kinspeople may live in a small grouping of one to three houses (*anda*), but most people live far from their neighbors. In some areas, for instance the hamlet of Kiya, where Peter lives, no one resides within 60 or 70 meters of anyone else. In Kolatika, around the Seventh Day Adventist church, people live within 10 or 15 meters of each other, desiring to be close in a form of religious solidarity. It is not uncommon to walk for some time

Figures 3.6 and 3.7. Social life in the *ama*.

through the forest, slogging through streams and mud holes, over logs, and up and down hills, and suddenly come upon a lonely house next to a large garden with a family living far from everyone. For some Porgerans this existence is much too "bush," however. With the development of the highway, village-like clusters have developed along the road.

Until fairly recently men and women lived separately due to male concerns over the polluting aspects of menstruating women (see

Biersack 2001; Meggitt 1964). A group of males related through blood and marriage lived collectively in men's houses (akali anda). Inside myths were recited and stories were told of past wars, wealth exchanges, and other socially important events. Plans for future exchanges and fights were also made. Surrounding the men's house were smaller houses occupied by individual wives, their children, and the husband's pigs, which were cared for by the women. At about the age of eight or ten, young boys were encouraged by their fathers and male relatives to live in the men's house. With the advent of Christianity in the 1960s, missionaries encouraged men to live in one household with their wives. Many men have done so, although husbands and wives still sleep in separate rooms.

On every group's andakama land there is a particular pool or small lake where the collectivity of deceased ancestors lives (as in the Mt. Paiam case above). When people die, they are called talepa (ghosts) and are believed to roam about within the group's land boundaries, causing misfortune among the living (Gibbs 1975: 38–39). When people fall sick, the first thing they do is make a sacrificial offering of a pig to the ghost of their deceased mother or father, or one of their grandparents if their parents are still alive. The anthropologist-priest Philip Gibbs surveyed Ipili attitudes about illness and sacrifice in the 1970s and found that 58 percent made propitiatory offerings to their deceased parents and 20 percent made offerings to their grandparents and generations beyond when sick (33). After a few generations individual talepa join the collectivity of deceased ancestors that is referred to as yama (spirits). The pool that contains the ancestral yama is called ipa ne, "the water that eats." The ipa ne is also a source for life-giving powers. Although unlike individual healing rituals wherein people sacrifice to an immediate relative, the sacrificial rituals associated with the pools are said to work on the health of the group as well as to renew the environment.

The way this works is as follows: The land in Porgera is conceptualized as being alive since it has a skin (yu umbuaini) and a bone (yu kulini). Individual people are said to hold (mina) the land skin, while the subclan holds (or owns) the land bone. Topsoil is associated with the land's skin, and the work that goes into farming it and growing crops in it is said to deplete the grease in the land (yu ipane). Rituals for replenishing the yu ipane involve the killing of pigs and carrying bundles of pig grease (yia ipane) up to the ancestral pools. The bundles are thrown into the pools to feed the ancestors, who ensure that the yia ipane flows back into the land to become yu ipane or soil grease. There is a cyclical logic to the concept of grease (ipane) and its role in the universe. The soil has grease, yu ipane, which is transmitted to the crops that Porgerans grow.

Work as farming converts the yu ipane to ipane that can be consumed as food. The food that humans grow also is fed to domesticated pigs, which are primarily used exclusively for rituals and exchange purposes, and ipane goes into both people and pigs through food consumption. The ipane that keeps people young, active, and alive is expended, however, through the work of production and reproduction. In reproduction the ipane of the parent flows into the offspring (breast milk is *andu ipane*, "breast grease," semen is *ui ipane*, "penis grease"). When these children grow up and their parents die, the ipane from production (gardening and pig rearing) is given back to the ancestors, who ensure that it flows back into the land to continue the process. A key point, however, is that the system of ipane is not self-regulating just through work but needs ritual intervention to continually build back up the life force in the universe that is lost incrementally.

The demise of these rituals (excepting the Paiam case) is the cause of much concern over the current fecundity of the land. Missionaries told Porgerans to "stop praying to Satan in the ground" and to instead pray to God in the sky. The removal of thousands of metric tons of land every day through mining also inspires worry about the effects on the grease in the land. Crops are said to be smaller, the land is perceived to be drying up, and there is an imbalance in ecological processes that cannot be resolved by local gardening practices alone. The efficacy of prayer to a Christian God is widely debated; many Seventh Day Adventists believe that God can improve soil fertility, while other denominations doubt this. One man, a Lutheran, captured eloquently what many others said to me: "I'm just one man living in a tiny country a long ways away from anything. Why would God care how my garden grows?" Nor does appeal to a Christian God resonate with the more than 30 percent of the people who don't identify as Christian.

Yu Nena: *Land of the Earth Mother Women*

"Cassowaries don't mate like other birds," Epe told me in 2006, when we were talking about species in the lower rain forest. "They conceive by jumping up and pounding their feet against the buttress of a *pipi* tree [*Syzygium* sp.]. The pipi is the father of the cassowaries. When the mother goes off to get food, she leaves the chicks with the pipi and it takes care of the young." For all of their detailed environmental knowledge, statements such as this one always confounded me about the extent and depth of people's ethnoecological knowledge. Epe went on to tell me that he wasn't sure if this was true or not now. "In the Bible it says that all animals mate, so maybe this story isn't true, but it's what all the elders say." Rather than an insight into the veracity of local environmental knowledge, however, being told about cassowaries mating with trees highlights the gaps in the relationships that Porgerans have with certain landscapes. Blindsnakes (*Typhlopidae*) turning into pythons, one species of marsupial being the juvenile form of a completely different species, and other such stories are illustrative of the disjunctive and ambiguous relationships that people have to the low-altitude rain forest where these species are found.

The interactions that people have with the low-altitude rain forest, or wapi, located roughly below elevations of about 1,600 meters above sea level, stem from their transitory experiences. Numerous resources are gathered in this warm, lush tropical rain forest, especially bird feathers (from lories, cockatoos, and various birds of paradise), cassowaries, and eggs from the incubating nest mounds of megapodes (*Megapodiidae*). Wild pigs, mushrooms and tree fungi, and *apate* fruit pandanus (*Pandanus conoideus*) are also found in abundance here and utilized by Porgerans. A critical constraining factor for frequent use, however, is the presence of malaria in the lower rain forest. Porgerans believe malaria has multiple causes, only one of which is transmission by mosquitoes. The most common source of malaria (and any other disease one may encounter in the wapi) is a female forest spirit called *yu angini wanda*, literally "earth mother woman." In Tok Pisin these spirits are called *masalai meri* (spirit women) or *wel meri* (wild, i.e., "uncontrollable" women), and evoke the greatest fear of any spirits in the pantheon of forest spirits that Porgerans discuss.

Similar to the role of the sky people in the upper rain forest, the spirit women of the lower rain forest are the caretakers of the flora and fauna of this region. Game and forest products are respectively referred to as the "pigs" and "sweet potato" of the yu angini wanda. Occasional forays into the lower rain forest to gather the pigs and sweet potato of the earth mother women are common in Porgera, but more frequent incursions into this area increase the likelihood of encountering one of these spirits. Great care is taken to avoid these encounters; men describe the approach of the spirit as the sound of a tree branch breaking so loudly as to sound like a gun. Sometimes the cries of the spirit babies they carry in net bags on their back can also be heard during their approach. Having contact with a spirit woman of the lower rain forest in its most serious form can cause death; most often, however, contact causes illness or confuses the senses so that people get lost in the forest. In most cases people are unaware that they have encountered a spirit woman and find out only later, when they defecate blood, as spirit women attack people by pulling out their innards and eating them without the victim's noticing.

Men are by far the most common travelers into the lower rain forest. Because women usually are with their babies, they are prohibited from bringing infants into this area, for fear that an earth mother woman will steal the child to keep for herself. Occasionally hunters are singled out by earth mother women for purposes of reproduction (which is why there are not spirit men in this zone of the rain forest). These spirit women will disguise themselves as a beautiful young woman whom the hunter has had his eye on for some time. She seduces him and they make love, allowing the earth mother woman to become pregnant. Afterward, as she walks away, he notices that she has ferns and moss growing out of her back instead of skin (see Gibbs's description of *yutuwana* in Paiela [1975: 47]). In reciprocation of a man's sexual services, the earth mother woman now sends him wild game (her "pigs") in abundance. According to Porgerans, even if one of these men hits an animal in the foot with an arrow, it will drop down dead. After some months of incredibly success-ful hunting, the hunter has a dream in which the earth mother woman appears. She tells him that it is now his turn to reciprocate the gift of wild game she has been sending him. She typically asks for the life of one of the man's human children. If he refuses, she asks for the life of a brother, wife, parent, or other relative. If he still refuses, she takes the life of her hunter-lover. Mangi Mole of Kolatika related a story describing human relations with earth mother women:

> There was once a man living down at Taipoko named Yewan who was married to an earth mother woman and had two children with her. I saw the woman and the two kids one time when I went hunting marsupials in the wapi with Yewan, a man named Koiya, and a Hewa boy who Yewan knew.
>
> We left Taipoko with our net bags full of sweet potatoes and started hunting. We hunted all day, but saw no game; although we did find ten fruit pandanus. Yewan said he thought he had seen some mar-supials back at one place he had passed earlier and told us to go cut wood and get the pandanus and sweet potato ready to *mumu* [cook in an earth oven].
>
> Earth mother women look after the marsupials, cassowaries, and wild pigs; the one that Yewan was married to would always give him game. Late in the afternoon Yewan came into camp carrying eight marsupials. We put them in the mumu and then went to sleep. The

next morning, Yewan told the Hewa boy and me to stay at the camp and watch the food, but not to eat any of it. He and Koiya went out to hunt some more. I didn't talk to the Hewa boy. We didn't know each other's languages, so we just spoke by hand signs. We sat around and then later started to gather some wild greens to mumu with whatever game they brought back. In the afternoon they returned with more marsupials, a cassowary, and a python. They were bent over from carrying so much meat. We cooked it all and then ate and slept.

That night, Yewan slept on one side of the fire by himself; the three of us slept on the other side. In the middle of the night I woke up from the cold. I saw a woman sitting next to the fire. Her skin and face were like a white woman's. Her two children were white, too. She was sitting near Yewan's legs; one child was by his waist, and the other by his chest. I acted like I was asleep, but I secretly spied on them to see what they would do, but I fell asleep and didn't wake up until the morning. When the sun came up, I looked over at Yewan and saw that he looked like he had been beaten up. He had a black eye and his nose was bleeding. He told us that the earth mother woman had come in the middle of the night and asked him for something to eat since she always gave him so much game. She had asked for one of us three to eat. He had told her he couldn't as we were his brothers [classificatory]. They fought then, and she had beaten up Yewan. We left and went back to Taipoko. Yewan walked the Hewa boy back to Hewa [across the Lagaip River].

A couple of days later, Yewan's son and Koiya's son went hunting in the wapi and the earth mother woman "ate" Koiya's son and he died when he returned to Taipoko. Koiya's wife was from the Piandani clan, and so was my mother, so I told Yewan he would have to compensate us Piandani. He said he wouldn't, so I came back to Kolatika and got some more men with Piandani connections and we went to Taipoko and seized ten of their pigs. They wanted to fight with us, but we told them to make compensation. They eventually did. Yewan is no longer alive. I don't know if his earth mother woman wife killed him or not.

Mangi describes the earth mother woman and her children as white. Other people said that in the past earth mother women were light-skinned black women, but in the past few decades their appearance has become more like that of white women with light-colored skin and long straight hair. This transformation has obvious parallels with the predatory appetite of Western culture that Porgerans have experienced from mining and consumerism. In local terms the relationship between hunters and earth mother women are expressed as marriage (ke la). But unlike the complementary and life-giving qualities of human marriage, marriage with spirits disrupts the social order and threatens hunters, their families, and their communities. In human marriage men give bridewealth pigs to women, who in reciprocity make gardens, raise pigs, and bear children, thereby reproducing society (Biersack 1995a). In an inversion of this, earth mother women give "pigs" to men who are expected to reciprocate with the lives of their kin.

Consequently Porgerans have a great fear of the lower rain forest. Unlike the positive labor that gardening and reproduction in the andakama produce in the form of food and children, the wapi carries aspects of negative labor in which food is received without working for it and society suffers as a result. This concern with spirit attacks in the lower rain forest also aids in explaining the limited environmental knowledge that some Porgerans have with animals in this area. The lives of cassowaries remain secret because to approach a cassowary and kill it is to risk encountering the society-threatening spirit woman who cares for it (cf. Bulmer 1967). In the past, it was only big men who dared to raise and exchange cassowaries. They had the wealth and knowledge to buy and know magic formulae (kamo) that could protect them from harm.

My own encounter with an earth mother woman took place in 1999. I was staying at the house of a man named Saima, a kinsman of Peter, in the Wailya (western Enga) area and heard the story of my close call only a few days after the event. I was staying at Saima's for a few weeks gathering some oral histories of the precolonial salt trade between Wailya and Porgera and of contemporary interactions between the two places. Initially Saima and his wife were happy to have Peter and me staying with them for, as Saima said, "there has never been a white person in this area before." My novelty resulted in Saima's house being packed every night while I learned about salt springs, origin myths, and other environmental resource management practices, and I talked to them about global happenings, such as the impending launch of the International Space Station (see Jacka 2005). Unbeknownst to me, these nights became progressively more stressful for Saima as he and everyone else started to hear the unmistakable sounds of the presence of yu angini wanda outside. Peter later told me that the spirit woman had heard that there was a "white skin" living in Wailya and wanted to come see for herself since she too had white skin. Saima secretly urged Peter to hurry up with our research as he was worried about the consequences of having an earth mother woman roaming around his house. My last night at Saima's I went outside to relieve myself, heard a funny noise in the forest, and came back in to see everyone's ashen faces. At the time I thought nothing of it and went to bed. I woke up the next morning with an excruciating headache and a high fever. Worried that I might have malaria, Peter, Saima's brother, Masmak, and I set off for Porgera, for if I was seriously ill, Porgera had a hospital. By the next day I was feeling better, and Peter and Masmak told me that they were worried because everyone had heard me get "attacked" by an earth mother woman while I was peeing, and they thought that I was going to die. I pointed out that Saima's house was in the andakama, not the wapi, so the strange noise we heard the other night couldn't have been a yu angini wanda. Peter and Masmak insisted that was why Saima and people living around him were so upset: a spirit of the lower rain forest had crossed a formerly impermeable boundary. This disruption of spatial norms is said to be a more frequent occurrence associated with the coming of whites and the changes in the land.

Another aspect of the wapi that causes consternation has to do with the waterways of the lower rain forest that carry the spirits of the dead to a mountain in the lower Porgera valley, where the souls (tandini) reside. As mentioned, after a person dies, his or her ghost roams around the clan lands attacking and "eating" kin. After a generation or so, the ghost joins the collectivity of deceased clan members (yama) in the clan pool, the ipa ne (the water that eats). Then a person's soul (tandini) travels in the waterways to dwell in the lower rain forest. At the junction of the Silia and Porgera rivers is a distinctive, pyramid-shaped mountain named Kewalieme where the souls of the dead dwell.[3] Hunters say that if you sleep too close to the Silia River you can hear the screams of the souls as they are floating toward the mountain.

The wapi is also associated with the Hewa culture group who live just north of the Porgera valley across the Lagaip River. The Hewa are believed to continually send sorcery to the Porgeran andakama in the form of various spirits that attack and kill people in the night. Owls and other night birds are the emissaries of the Hewa, who are constantly probing for unsuspecting victims. This flow of black magic uphill inspired one man to respond to my question of why men's houses are always built on the tops of hills by saying, "We do this to put the women's houses between us and the Hewa sorcery."

That Porgerans conceptualize the three altitudinal zones as disparate is also encoded in how they are obligated to speak. In the lower and upper rain forests, the words for domestic household items (axes, net bags, etc.) cannot be spoken, nor can the words be used for trees and food crops grown in the andakama, the domestic zone between the upper and lower forests. Those who break this taboo (*laka la*) risk twisting an ankle, cutting themselves, falling off a cliff, getting lost, not finding whatever it is they came to hunt or gather, having their dog run away, getting rained on, and so forth; essentially they risk spoiling their trip and, at worst, dying. The following list describes some of the transformations of everyday objects in the language of the forest (cf. Bulmer's discussion of "pandanus language" among the Kalam [1967: 15n18]):

sweet potatoes (*ulia*):	*ita pini*, "tree roots"
a net bag (*nu*):	*mandi mandi*, "that which carries"
taro (*ma*):	*asiki*, "wild taro"
fire tongs (*petoko*):	*andaka piti piti*, "that which sits in the house"
bow (*yandate*):	*mina mina*, "that which is held"
salt (*ipi*):	*peyatau*, "shake it"

All marsupials must be called by their generic name (*koa*), as must trees (*ita*), rather than using the appropriate species terms; likewise all the various species and varieties of pandanus are called by their generic name (*anga*). While I am unwilling to argue that the andakama, the middle zone of houses and gardens, is a space of domestication as opposed to the "wild" spaces of the upper and lower rain forest, the proscriptions around behavior and language use signify the way that these ecozones are perceived by Porgerans as unique.

One of the key goals of this chapter was to address the relationship between the materiality of everyday livelihood decisions and the perception of spiritual landscapes. As discussed earlier, the total environment plays a critical role in shaping human affairs in the Porgera valley. Materiality and perception work in unison and are themselves co-constituted. In Porgera a dynamic environment is envisioned as containing powerful spiritual forces that regulate and influence ecological processes. Porgerans attempt to capture these creative and regulatory forces through prosaic activities like gardening, marriage, and exchange as well as more esoteric means of ritual and spirit propitiation.

CHAPTER 4

People: *Wandakali*

The Nature of Production

This chapter examines the interrelations between human production and the biophysical environment to illustrate how Porgerans conceive of human agency as essential for producing both nature and society through work and exchange. A basic theme that runs through this chapter and is picked up again in chapter 7 is the fundamental importance of work (*piape*) for Porgerans in constituting society and the environment.[1] I also demonstrate the ways that Porgerans "work" to create social groups that ensure flexibility in terms of group composition and land-tenure dynamics. Many of the transformations in Porgeran ideas about their mutually constituted social and natural world is a product of the ways that colonialism, mining development, and Christianity have altered their ideas about work and the nature of production. As Marx pointed out, different modes of production engage with the natural world in different ways, thereby organizing the forces and relations of production differentially as well. Many environmental anthropologists following this trend actively seek to understand the particular ways that capitalism organizes human-environmental relations. Neil Smith (1996: 46–48) claims that in some of this work there is a "domination of nature" narrative that sees the technological apparatuses of capitalism as especially successful at subjugating and degrading the natural world

(Merchant 1990). Smith critiques this viewpoint on two fronts: the first is that it posits an autonomous nature that is an external thing-in-itself separate from humanity; the second is that domination implies control. According to Smith, Marx was right in saying that there is something central about labor's relationship to nature, but what we need to focus on is the "production of nature" (Smith 1996: 49). The production of nature "incorporates material with conceptual construction" and "is as much a cultural as it is an economic process and should be understood in the broadest sense of transforming received natures" (50). Moreover, if we focus on the idea of production rather than control, we are better able to understand how different scales of production constitute the environment in alternative ways. I must stress, though, that my argument is not that nature is a social construct (Bakker and Bridge 2006) but rather that Porgerans understand the fertility of soil and natural processes as dependent on reciprocal relations between humans and nonhuman spirits.

One of the critical insights that political ecology brings to human-environmental studies is the argument that Malthusian ideas about population and land degradation are flawed and that land degradation is not solely an ecological problem (Blaikie and Brookfield 1987b). John Bellamy Foster (2000: 95) describes Malthus's notion of the relationship between human production and natural soil conditions: "Along with his great contemporary, David Ricardo, Malthus saw the fertility of the soil as subject to only very limited improvement. Nor was soil degradation an issue. . . . For Malthus, the properties of the soil were not subject to historical change, but were simply 'gifts of nature to man' and, as Ricardo said, 'indestructible.'" Porgerans, as farmers, would scoff at Malthus and Ricardo, claiming that soil fertility is inextricably linked to historical change and human interaction. And not only is soil fertility dependent on human labor, but the very process of horticultural production is under threat from an invasive forest that labor keeps at bay.

An example of what I mean may be illustrative. Peter had a maternal uncle who died in 1999. After hearing of his death, we walked out to the hamlet where he had passed away to see his widow and sons and pay our respects. Most Porgerans consider Taipoko, where the family lived, to be a remote bush place, and in fact Peter's uncle's immediate kin were the only remaining people still living and gardening there. To reach Taipoko it took us several hours of walking through the dense rain forest along trails with knee- and thigh-deep mud bogs (figure 4.1). A flock of sulphur-crested cockatoos flew from treetop to treetop keeping pace just ahead of us, screaming their shrill calls back and forth. Once

Figure 4.1. The author on the trail to Taipoko.

every hour or so we would break out of the forest into an abandoned garden, glad to be back in the bright sunlight. With the exception of these few clearings, it was just tall trees, shrubs, and climbing bamboo everywhere we looked. When we arrived Peter lamented that with his uncle now dead, the rest of the family would probably move closer to the highway to escape the bush: "All of his work, his gardens, his pandanus trees, the forest will come and cover these up now." At the heart of Peter's concerns was the continual threat that Porgerans live with, which is that their life's work will succumb to the forest.

Popular imagery of rain forest dwellers often depict indigenous people referring to the forest as a "mother" or a "protector" (Bird-David 1990; Brosius 1997). This is not an image that Porgerans would use to describe their forests. As I detailed in chapter 3, humans use forest products only after engaging in relations of reciprocity with potentially harmful and dangerous spirits. As a consequence Porgerans are not natural conservationists, nor are they naturally destructive of the environment. There are in reality far more complex factors going on that cannot be reduced to easy formulations that fit into Western categories deployed by environmentalists and their critics. (For contemporary critiques in PNG, see Kirsch 2008; Macintyre and Foale 2004.) In her book *Friction*, Anna Tsing (2005) argues that among the Meratus in South Kalimantan, Indonesia, forests are best understood as social forests. Tsing describes secondary forests as occupying "gaps" between the visible realms of agriculture and old-growth forest; these gaps remain unseen by developers and government agents. She asks, "How might we learn something from the landscape of regrowing forest? Again it makes sense to ask how its Meratus inhabitants know it. Here the key is that secondary forest is always a social place; to know it is to know the history of its flora and fauna in relation to socially situated human biographies" (190). This is certainly the case in Porgera as well.

For Porgerans the natural world by itself is a mostly unproductive yet agentive environment that actively erases the human impact upon it. Human labor—cutting down trees, building fences, digging ditches, and planting gardens—makes the land useful (see Bulmer 1982 for Papua New Guinea; Haenn 1999 for a similar example from Mexico). In 2006, as I was standing with a group of men looking at the mining waste dump along the Porgera River, one man spoke out: "All of this land used to be nothing. It was just forest. Then I came and fenced it, dug ditches, and made gardens. I planted flowers, made paths, and planted pandanus and casuarina trees. Now it is good land." The men complained that the mining dump, which was covering up their land, was now erasing their years of labor to convert "nothing" (i.e., forest) to "good land" (i.e., gardens). Porgeran action on the landscape, the products therein (from gardens and agroforestry plots), and even the landscape are inscribed by material and symbolic practices. They are material in that they reflect the varied processes of human activity on the natural world. They are symbolic through the cultural ideals that are encoded in the landscape as people labor in their gardens. Central to these material and symbolic practices is the notion that, ideally, productive work is gendered. Men cut forest, fence gardens, and dig ditches. Women plant and harvest gardens. The very nature of productive activity in Porgera captures the tensions in cultural expectations of sexual separateness and complementarity with ideals of women's subordination. On the one hand, the Ipili term for "people," *wandakali*, is composed of the words *wanda* (woman/women) and *akali* (man/men), or people as women-men. On the other hand, gardens embody cultural norms of subordination in that women should stay fenced "in the middle of the garden" (Biersack 1995a; Wardlow 2006). This they do, as the never-ending task of harvesting sweet potatoes and making new mounds occupies a vast amount of their time. For both women and men, though, people say "Ke ko, ki epene," which translates literally as "Legs bad, hands good," meaning that the activity of hands (productive work) is good, while the activity of legs (ambling about and getting into trouble) is bad. Numerous times while I was walking through

Figure 4.2. Muyu with his grandchildren at Yakati Ipati.

the forest with a group of people we would come out into a large garden clearing. The mounds of sweet potatoes, bunches of sugarcane, edible greens, pandanus trees, banana plants, and other vegetables would always elicit whistles of appreciation and the tsk-tsking along with the side-to-side head shaking that Porgerans do to express astonishment. "Man, look at the size of this garden," people would say in awe of the human effort it took to create. Once while standing in the midst of a ten-hectare clear-cut that was part of a coffee development project in Porgera, I remarked to Peter and Ben how sad it was to see so much primary rain forest cut down. They both looked at me disdainfully. Ben laughed in astonishment and disbelief and said, "What are you talking about? Can't you imagine the size of the garden that could go here?"

Peter's family was renowned for having some of the best garden land in the eastern Porgera valley. The garden area, called Yakati Ipati (figure 4.2), illustrates how important work and ancestral relations are for subsistence in this region. Yakati Ipati, literally "the wet muddy area where the Highland breadfruit tree (*Ficus dammaropsis*) grows," contained the area's most prized soil, *yu kunguluma*, black soil that is not too clayey and is friable. Muyu and his children had been gardening continuously here for as long as he (then in his eighties) could remember. Yakati Ipati, along with a few other gardens in the area, was famous for not having to lie fallow like most of the surrounding land. (For other examples from neighboring groups, see Sillitoe 1996: 36–39; Waddell 1972: 44.) Muyu's father, Yakati, named after the tree (a common practice in Porgera), had long before cut much of the land at Yakati Ipati by himself out of primary forest (*itapu*) and gardened on it, eventually passing the land down to Muyu and Muyu's sister, Aiyo, who then designated portions of it to their children. Yakati's clan brother Wuambo was, according to his son, Kipan, a "lazy" man who refused to cut primary forest and instead spent his time trying to weasel his way onto Yakati's garden land. Eventually Yakati's line of descendants pushed Wuambo's descendants off the garden lands that Yakati had originally cut, leaving men like Kipan with no alternative but to cut primary forest to make their own gardens, a task few men are willing to undertake due to the labor involved in cutting, fencing, and digging ditches (see Sillitoe 2010).

Whereas the *ama* is the focus of social life in Porgera, the garden is the center of household family life. Most gardening activities are gendered: men cut the trees and shrubs where the garden will be located, build fences around it to keep out wild and domesticated pigs, and dig ditches to drain off the excess water from rainfall; women plant and weed the gardens and generally do most of the harvesting. (For an overview of the rich literature on horticulture and environmental relations in PNG, see Allen et al. 2001; Ballard 1995; Bayliss-Smith 1996; Bourke and Harwood 2009; Brookfield and Brown 1963; Clarke 1971; Grossman 1984; Hviding and Bayliss-Smith 2000; Modjeska 1982; Rappaport 1971, 1984; Sillitoe 1983, 1996, 2010; Sillitoe and Strathern 2002; Waddell 1972.) The word for "garden" in Ipili is *e*. Porgerans distinguish gardens by whether they are new (*e wene*, "new garden"), currently producing (*e siaka*, "alive garden," or *e talu*, "fertile garden"), fallowing with grass (*e si*), or fallowing with trees (*e pale*). After about fifteen to twenty years, *e pale* become *e imati* (forest-cut gardens), which eventually revert back to forest. Forests, *itapu* in Porgera, *siapu* in Paiela,[2] are distinguished by whether they are forest preserves (*ima*), or just plain forest (*itapu*). I translate ima as "forest preserve" as there is some sense that ima lands should be protected; people are allowed to harvest timber for personal use from them but are not allowed to clear-cut them for garden and house sites. Most men prefer to cut gardens in e pale, second-growth forest,[3] that was originally gardened by one of their ancestors, or themselves if they are older (cf. Sillitoe 2010: 160). If certain hardwood tree species are used for fencing, much of the original fence may still be usable fifteen to twenty years later. Often the ditches require a bit of cleaning, but in a very short time a piece of land is ready to be planted. *Yawale* (*Casuarina oligodon*) trees and pandanus trees mark the sites of e pale. Casuarina and pandanus

trees are individually owned in Porgeran society and serve as mnemonic
markers for garden lands that one's descendants are given the first
claim to garden. In fact planting casuarina and pandanus trees is the
mechanism for claiming land in Porgera. Consequently land disputes
frequently break out when landowners (yu anduane) find out that people
with limited usufruct rights only (i.e., epo atene, "those who come and
stay," a.k.a. immigrants; or yu nolo atene, "those who come and eat the
land," a local person possessing only use rights) have planted trees on
land loaned to them.

Once a man has cleared, fenced, and ditched the land, then the woman,
usually accompanied by her daughters, turns the soil with a small-bladed
shovel; traditionally she would have used a digging stick. The women
mound the soil over any remaining weeds and shrubs and sometimes
add the leaves of pandanus to the inner mound. The decomposition
of the inner mound fertilizes the sweet potatoes and other vegetables
planted in the mounds. Unlike many fallow systems, highlands New
Guinea mounding actually allows for long-term in situ gardening and
does not require the large amounts of land that swidden farming does
(Sillitoe 1996). In older forest, mounds (called alo) are smaller, approximately 0.25 meter high and 1 meter wide, and often are planted only with
sweet potatoes. Mounds (mondo) in fallowed gardens are over 0.5 meter
high and about 2 meters wide and are planted with sweet potatoes and
other vegetables. A freshly mounded garden is an amazing sight—not a
living plant is left in the ground, just mounds of dirt with sweet potato
slips. Within a month beans, onions, greens, tomatoes, corn, soybeans,
and other vegetables cover the mounds, and the sweet potato runners
have covered almost the entire garden. Within two months everything
is harvested except for the sweet potatoes, which mature in four to ten
months, and the sugarcane, which grows for a year or two (figure 4.3).

The lack of seasons in Porgera means that women work continuously
planting and harvesting sweet potatoes. The tubers do not store well,
and fresh ones are needed daily. Every afternoon women return from
the gardens with big net bags of sweet potatoes strapped over their
foreheads and hanging over their backs. Infants ride piggyback style
on the mother's shoulders. The daughters each have their smaller net
bags of produce and carry the shovels. The heavier rains of November
to March slow tuber production but do not curtail people's gardening
activities. As mentioned, pandanus trees peak in their fruiting during
this time, helping to offset the loss of sweet potatoes.

The work of gardening illustrates gender complementarity and reciprocity between the sexes. The man's opening "gift" of cleared and fenced

Figure 4.3. Gardens in various states of production. Pandanus and casuarina trees mark the borders.

land to his wife is reciprocated by her labor in producing sweet potatoes that feed the family and the pig herd. The surplus pigs that a woman's labor produces are channeled into exchange activities that bring renown to the man and, in Porgeran culture, his wife. Ipili speakers also say that children are a return "gift" to a man from a woman in recognition of the bridewealth that his clan paid to her kinspeople (Biersack 1995a: 241–43). The logic of reciprocity also governs human relations with the spirits of the andakama, namely the clan's deceased ancestors. As Porgerans say, "Gardening is hard work." As discussed in chapter 3, living humans give gifts of pig grease (yia ipane) to deceased clan ancestors, who then replenish the grease in the land (yu ipane), allowing crops to flourish, which are then consumed by the people and pigs. Gardening in this sense is not just growing food; it is the central process in regulating reciprocal relations between men and women, people and pigs, pigs and ancestral spirits, and all of these entities with the land.

The Production of Society

Just as "unproductive" forest requires human agency to make it into productive garden, humans are also essential for making people productive through puberty rituals and through the process of creating and maintaining social networks and groups. Puberty rituals for women were performed individually; young women would go out to the cane grass and shake morning dew on their skin to grow their breasts and maintain healthy skin. I am not sure whether young women continue to do any of these rituals today. In the Porgera valley there were two male puberty rituals that were done in groups led by senior bachelors. The first of these was called *omatisia* and was a ritual that young men did to purify and grow their body and which centered around the ritual marriage of a sky woman associated with the upper rain forest (Biersack 1998a, 2001, 2004). Another part of the omatisia ritual was for men to grow their hair long, a sign of male health and vitality. After growing out their hair for some months, they would cut it and save it to make the bullock-horn wigs that Porgerans were renowned for and which are still worn during important exchange and ritual ceremonies. The other collective male puberty ritual done in the eastern Porgera valley was the *sanggai* that came from the Enga in the early twentieth century. This ritual too involved purifying and growing young men's bodies along with symbolic marriage to a spirit woman (see Meggitt 1964; Wiessner and Tumu 1998: 215–44). While none of the male puberty rituals are still performed in Porgera, I mention them to highlight the importance that human agency has in "growing" young people into adults, which is similar to the horticultural work of gardening.

Today people are more likely to claim responsibility for helping to "grow" a young person into an adult through their efforts in child care, paying school fees, and purchasing commodities. (For ideas about Melanesian sociality, see Sillitoe 2010; Strathern 1988.) As an example, in 1999 I went to a court case over the bridewealth compensation of a woman named Pimbala September.[4] Her father was from the Pakoa clan but married a woman whose father was from the Nomali clan and whose mother was from the Tokoyela clan. He and his wife spent most of their time living on Tokoyela lands. He was killed in 1988 as a payback killing for an Undiki man that the Tokoyela had killed during earlier tribal fighting. In 1999, when Pimbala got married, her bridewealth payment was fourteen pigs and K1,000, which her Nomali relatives claimed during the bridewealth compensation. The Pakoa clan (her father's paternal clan) initiated the subsequent court case, claiming that the majority of the bridewealth payment should have gone to them. The village court magistrate told the Nomali to pay the Pakoa claimants four pigs and K400. To prevent any hard feelings the Nomali gave more than requested: five pigs and K500. During the court proceedings the Tokoyela vociferously complained that they too should receive some of the bridewealth payment, as it was they who actually cared for Pimbala after her father was killed (Tokoyela being her mother's maternal clan). Peter confronted his Pakoa cousin, Yumu Mangala, and cried out, "We [Tokoyela] bought her soap and clothes after September was killed. We paid her school fees. We grew her into the woman she is now. It's wrong for you to take all of the pigs." Shouting broke out between the Pakoa and Tokoyela, and Tokoyela men started forcibly taking the pig ropes from Pakoa men. A lineage brother of Peter's brought out a homemade shotgun from his nearby trade store and threatened to kill any Pakoa men who tried to leave with the "Tokoyela's" pigs. Accustomed to Porgeran compensations, once the yelling started I had already moved to the outskirts of the proceedings, crouching behind a boulder while keeping watch over the events. Eventually the tension dissipated after the Pakoa agreed to turn over some of the pigs to the Tokoyela.

Allied to the process of making humans, Porgerans refer to the process of organizing marriages and conducting marriage compensations as *yame pi*, "making clans" (Biersack 1996a). Biersack argues, "People should marry strategically, so as to replicate the conjugal decisions of their consanguines" (21). The importance of marrying strategically is to ensure access to a variety of productive garden lands, as it is primarily through kinship that a person makes claims to land. The way that making clans ideally operates in the Porgera valley is that people recognize themselves as having membership with the eight clans of their great-grandparents: four from the father's side and four from the mother's side. Anyone who also has one of these clans in any of his or her eight clans is considered a kinsperson and unmarriageable. How, then, do people marry strategically and replicate the conjugal decisions of their ancestors? I present two examples of how this works, but first I refine what I mean by the term *clan*. In the eastern Porgera valley, people generally refer to a named group that owns (holds) a particular block of land as a *tata* (clan; *yame* in western Porgera and Paiela). As one man described to me, "Tata is like the trunk of a tree." All clans have smaller groups within them (subclans) that are called *tata anako* (small tata). The tata anako are "the branches of the tree." Subclans are made up of smaller groups (called *yumbane* or tata anako), which roughly correspond to lineages, "the leaves of the tree" in local parlance. Exchanges between groups occur at the tata anako level (either subclan or lineage), although people generally just use the tata name to indicate the exchange partners.

In the first example of strategic marriage, or what I refer to as "marriage clusters" (figure 4.4), Peter's father, Muyu, married Kainalem, a Pandame woman. Muyu was a Tokoyela-Maia-Itapu lineage member.

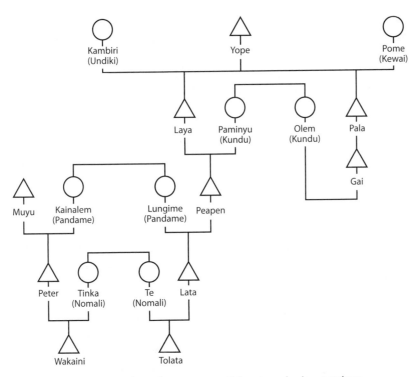

Figure 4.4. Marriage clusters among Tokoyela male clan members.

After marrying Kainalem, Muyu urged Peapen, a Tokoyela-Maia-Yope lineage member (a subclan "brother"), to marry Lungime, Kainalem's sister (a biological sister). When Peter got married, he of course couldn't replicate his father's conjugal decision to marry a woman from the Pandame clan as that was his mother's clan. He married a Nomali woman named Tinka. But Peter then worked to get Tinka's biological sister, Te, to marry Lata, Peapen's son, who was both his cousin through his mother and his "brother" due to shared paternal subclan membership. Peter's son, Wakaini, and Lata's son, Tolata, if not for the strategic marriage clusters pursued by their fathers and grandfathers, would of course have considered themselves kinsmen due to the fact that both were from Tokoyela-Maia subclan. But since Wakaini and Tolata also have mothers *and* grandmothers who are biological sisters, their kinship ties are even more dense and overlapping.

The second example again involves Peapen, who is the product of a marriage cluster arranged by Yope, the founder of the Tokoyela-Maia-Yope lineage. Yope had two wives, Pome, a Kewai woman, and Kambiri, an Undiki woman. With Pome, Yope had a son named Pala, who had a son named Gai (Ben's great-grandfather). With his second wife, Kambiri, Yope had a son named Laya (Peapen's father). Laya married Paminyu, a Huli woman from the Kundu clan. With Yope and Laya's sponsorship and urging, Gai married Olem, Paminyu's biological sister. While Gai and Laya were already paternally related, Laya was Gai's "little father," and by marrying sisters they became brother-in-laws as well. The sons of Gai and Laya, while already paternally related, were now maternally related too. As shown in figure 4.4, Peapen, Laya's son, went on to create his own marriage clusters with Muyu, his son Peter, and Peapen's son Lata.

Similar to the Huli (Ballard 1998), Porgeran kinship is structured by gender and precedence. There are a number of terms that Porgerans use to indicate how they are related to each of their clans. A man is considered to be a *tene* in his father's father's clan. *Tene* means "source, origin, base, reason" and in kinship means a descendant through two male links. Women's status is *wane* (female descendant). In regard to his mother's father's clan a man is called a *wanaini*, literally "woman's son"; this term indicates all the male members of the clan descended through a woman. The son of a wanaini is called *aiyane lapone*, literally "second father," which indicates descent through a woman two generations prior. This term is also used to refer to a man's father's mother's clan. The son of an aiyane lapone is called tene because by this time there are now two male generations through which the person can claim descent. And more than likely the original female member of the group is deceased from old age. A man often refers to his mother's mother's clan as *aputini*, "grandmother," or *angini lapone*, "second mother." Men who marry into a clan are called *wanakalini* (woman's husband) by that clan. And wives of clan members are called *akali wetene* (man's wife). All of this indicates that there is a slight patrilineal bias in this cognatic kin system. Over time cognates, descendants of females, are "incorporated" (Wohlt 1995) or turned into agnates, descendants of males.

All of this is important because people have different ways of being affiliated with clans that construe varying rights to land and expectations of membership. Having multiple clan affiliations also means that people are considered to be "between" (*tombene nga*) certain clans. This proved difficult for the Australian colonial government during censuses. In the first census in Porgera, in 1957, one patrol officer noted that "the only hitches occurred when some people presented themselves with other groups after their own sub-groups had been recorded" (Wabag PR #2, 1957/58). In the example regarding September, he was a tene in Pakoa and a wanakalini to both Nomali and Tokoyela, his wife's father's and mother's clans, respectively. My Tokoyela informants referred to him as a *tombene akali*, an "in-between man" or "middle man," who could serve to intervene between warring clans and oversee compensations between clans. Tragically his social position as a tombene akali failed to prevent his assassination when another clan wanted to strike at the Tokoyela, for, according to several people, he was singled out for his status as a leader rather than a tombene akali. One of the most significant roles of tombene people is to encourage and sponsor marriages that duplicate their own status between clans, as discussed earlier in reference to marriage clusters (see also Biersack 1995a, 1996a). The density of overlapping kin networks that this creates helps to promote peace in an environment in which endemic warfare is very high.

To continue with the social details surrounding September's assassination, the Tokoyela-Maia subclan and the Undiki-Kewane subclan started fighting in earnest after his death as one man had now been killed from each side. Three Undiki-Kewane men were killed who were also the sons of three brothers. These three brothers urged the Undiki to stop fighting and negotiate for compensation payments to begin, which would mark the end of warfare. They did, and Wamope Lenane explains, "When the time came for making compensations, the Undiki-Tapaiya, my *lain* [Tok Pisin for "clan" or "subclan"], came to help the Itapu [one of the lineages of the Tokoyela-Maia], the Undiki-Punuma came to help the Yope [another lineage of Tokoyela-Maia] pay pigs. The Itapu lain was related to Tapaiya through two mother links. Yope was related to Punuma through three mother links. The Undiki-Kewane told the Undiki-Tapaiya and the Undiki-Punuma that if they were going to help the Tokoyela with compensation pigs, then they should just go live with them" (field notes, February 1999). Wamope's statement demonstrates the importance of female links in this cognatic society even when female links are erased by the incorporation of outsiders. This recognition of women comes about through a saying that men use when making compensation to distant mother lines: "Ipapuni ombo eyama," literally "We cover the vines of the sweet potato," used metaphorically to mean that one must not forget one's ancestresses and these women's kin. Kinship is thus perceived vertically as descent through the symbol of trees (cf. Hagender 2000; Jones and Cloke 2002; Rival 1998)—trunks, branches, and leaves—and horizontally as marriage alliances through the imagery of sweet potato vines.

Leaders attempt to influence compensations and marriages so that alliances are built up over time and through generations of kinship. While there were a few Tokoyela and Undiki marriages in the past, there were never enough for the in-between men to prevent the fight in 1988. Ten years later there still had been no marriages between the two groups, but many men assured me that if someone were to initiate such a union there would be great support on both sides. As one man claimed, "Everyone would give lots of pigs to make sure that this marriage was strong." Big men also work to prevent support of unions that they feel bring nothing to the group. When Ben got married he said hardly anyone supported him with pigs and money. "I was marrying a Pilisame woman. Pilisame was a sister of Maia, and since I'm Tokoyela-Maia no one wanted to help me," he explained. (This was a particularly scandalous marriage as Porgeran subclans take women's names. Tokoyela-Pilisame is a landowning subclan in Tipinini.)

Malu and Multilocality

As mentioned, at birth Porgerans are members of the eight clans of their great-grandparents (for more detail, see Jacka 2003: 107–10). With their individual life choices, however, people develop stronger affiliations with some clans and allow others to wither. Generally people remain closely allied to their father's and mother's clans but also may develop ties with their spouse's clans or even join a group with whom they have no kinship relations. Maintaining and claiming membership in one's clans is "hard work" (like gardening), according to Porgerans. Not only must one contribute to group exchanges, build a house and garden, and affiliate with one's clan members, but one must also have knowledge of a clan's malu, or genealogical origin story.

Malu begin with a story in the distant past, when there was no separation between humans and the nonhuman world. For example, clan founders can be tree kangaroos (Tokoyela), pythons (Bipe), dogs (Yawanakali), even limestone (Piokai) and grass (Yandan). In all of these cases the clan founder marries and then bears offspring who, after a few generations, become the subclan founders. During the telling of malu, people move from an important story linking the clan founder to the current landscape and the recitation of genealogical linkages, beginning with the founder, moving through the founders of the subclans and lineages, and ending with the storyteller. All the malu I heard recited from ten to fifteen generations between the contemporary storyteller and the clan founder. Malu thus serve as origin stories about the division between culture and nature and also allow the teller to make claims to particular land areas. The Maiango clan is a mixed Ipili-Enga group with lands in Wailya, a day's walk north of the eastern Porgera valley. Semai Katokope, a Piokai (western Enga) clansman living among the Maiango, provided a detailed malu of Maiango (demonstrating his right to live there):

Maiango didn't have any parents. He was the first one. A *pai* [*Castanopsis acuminatissima*] tree gave birth to him. He lived at Alua Ipati [the swamp of the *alua* bird]. One day he put some salt and ginger in his net bag and traveled up to Polokombi [Wailya was the source of salt for Porgeran traders]. He followed the Diyu River and came up to Wauwusa. He then came up to Mt. Di. He rested at Mt. Di and before leaving took out his stone ax and cut the face of Mt. Di [there is still a mark there today]. Then he climbed up Mt. Tupalaku and then continued on to Kumamunga, where he saw a house. There was no one at the house, so he went inside to see who lived there. Next to the fire, instead of a stack of firewood there was a stack of dead cassowaries. He thought to himself, "Cassowaries are food, not firewood. Who is doing this?" He was hungry, so he decided to cook some of the cassowaries. He cut up a pile of wood, got a fire going, and started heating stones to use for a mumu. He plucked the cassowaries, gathered some wild greens, and started cooking everything in the mumu.

While he was sitting there waiting for the food to cook, two young women entered the house; their names were Ii and Aiya. Maiango was shocked because the women were completely naked and also were lacking vaginas. They asked where he had come from—he said Alua Ipati—and what he was doing. He said he was cooking the cassowaries; they were food, not firewood. He removed the cassowaries from the mumu and cut them up. He piled up meat and greens for each person and gave the women some salt and ginger. They all started eating. The two young women couldn't believe how good the cassowaries tasted. They told him that they would go out the next day and catch a bunch of cassowaries to eat tomorrow.

The next day the two women told Maiango to get everything ready for a mumu while they went to get the cassowaries. The two usually just went up to the cassowaries and grabbed them. Today, however, all the cassowaries were afraid of them and ran from them and called out to all of the other cassowaries to run away. The women chased and chased the cassowaries all day, their skin was scratched up from thorns, and they only came home with two cassowaries to eat. The women explained to Maiango that now all the cassowaries had become wild since they started eating them. Maiango said not to worry about it and told the women to come back and live with him at Alua Ipati.

The next morning the sisters gathered up all of their things and followed Maiango along the Wanyom River back to Alua Ipati. At Kuta Ipati one of the women fell, spilling sedge seeds that women use for grass skirts. Women still gather sedges for skirts at this swamp [*kuta* is the Enga word for "grass skirt"]. Arriving at Alua Ipati, while the women went into his house, Maiango took a sharp pig hoof, a broken piece of *kina* [mother-of-pearl] shell, and a piece of *atokoi* stone [a sharp stone used for cutting] and shoved them into the trunk of a banana tree. He then smeared pig grease all over the trunk. He called out for the sisters to come and climb into the tree to get some bananas. Each one tried to climb up but slipped on the greasy trunk. When the women slid down the trunk, the sharp things cut their vaginas open and they started to bleed. Maiango told the women to cover themselves with grass skirts and to learn shame.

These three are the ancestors of all the Maiango. Maiango "carried" Kengele and Minandiwa [in Ipili, Enga, and Tok Pisin, the verb "to carry" is used to indicate that someone is a direct descendant of a person]. Kengele and Minandiwa carried Tikipai and Atipai. These two carried Kilima and Laiyima. (field notes, June 1999)

At this point in the malu Semai begins to relate the wives' names and clans of Kilima and Laiyima and all of the children they gave birth to, continuing on for several more generations, ending with all of the children living on the Maiango's land today.

The Maiango malu is interesting on several levels. It relates the time that animals became wild, when humans start to eat them. In other people's versions of this malu, the three humans are also responsible for placing certain species of birds at particular altitudinal zones on the landscape and are the cause of various tree and plant species growing in an area. Maiango is not just the founder of a clan but is also responsible for the beginning of human reproduction by creating vaginas and menstruation in Ii and Aiya. The very items he used—a pig hoof and a kina shell—would become the critical exchange items that constitute bridewealth and marriage and all of the rights associated with them.

As Semai's detailed accounting of the Maiango malu demonstrates, knowing a clan's malu and having the ability to link oneself to that malu through kinship is the main means of making a claim to that clan's land. The goal in having affiliations with multiple clans is to live a multilocal lifestyle (Biersack 1996a), having gardens and houses on multiple clans' lands, assisting with exchange and warfare obligations with multiple groups, and allowing one's name to be known across the land.[5] This form of multilocal residence is called *pu ipu* in Ipili and literally means "go come" or "those who come and go." There are both ecological and social benefits to such a system. Ecologically Porgera suffers from poor, shaly, clay-based soils on steeply sloping ground. Spreading one's gardens around lessens the chance of crop failure that could detrimentally impact the household. Socially, having multiple affiliations provides refuge for the frequent outbreaks of interclan warfare that pervade Porgeran society. During wars individuals can decide to either stay and fight or pack up and move elsewhere, especially if the source of the hostilities involves someone in their group whom they don't like or with whom they recognize close kin relations.

The system of multiple clan affiliations, while providing an outlet for people fleeing interclan hostilities, also, paradoxically, generates a significant amount of intergroup hostility. In July 1999 my friend Kipan Wuambo told me that there was going to be a *laita* exchange held, where he was going to get a pig. A *laita* is a payment given by a woman's adult children to compensate her (and her kin) for her efforts in raising them. As Peter Enopi, the big man of the Makepo clan, explained to me en route to a different laita, "It cleans her babies' shit off of her hands." In the case of Kipan's laita, a man named Angalo was paying laita for his deceased brother, Bali, who had been killed in a tribal fight. Instead of giving laita to his four mother's clans, Angalo announced during village court hearings, a few days before the laita, that he would just give pigs to his *auwene* clan (mother's father's mother's clan; literally "ancestress's clan"), and not the other three. Kipan and Saima (whose house I stayed in when I encountered the earth mother woman) started arguing with Angalo, saying that as Kakayuni clan members they deserved to get the pigs as Kakayuni was Angalo's mother's "true" clan (her maternal clan; in Ipili, Angalo's *angini* [mother] clan). I was puzzled by their argument; I had always thought of Kipan as a Tokoyela and Saima as a Maiango, as they each lived on those respective clans' lands. During the announcement of the upcoming payment, Angalo refused to acknowledge Kipan's and Saima's claims to the compensation.

Later that evening I asked Kipan why he was trying to claim pigs as a Kakayuni, since in my mind he was really a Tokoyela. Kipan replied that long ago there were five Kakayuni sisters, one of whom never married and four who did. Over the next half hour or so, he described in excruciating detail all of the people who were descended from these four sisters. Their living descendants included people in several clans located in Tipinini, Kairik, and Wailya, including himself and Saima. He concluded by claiming, "We're Kakayuni. We should get some of those pigs."

True to his word, a few days later on the morning of the laita, Kipan had organized a group of "Kakayuni" men from around Tipinini, Kairik, and Wailya to march up to the Yomondaka ama grounds to claim some pigs from Angalo. Some of the Pakoa men joined the group as they were one of the other mother-side clans who weren't being compensated. Many of the men carried spears, and all were armed with axes and machetes. They sang and shuffled their way from Kolatika to the Yomondaka ama. Between refrains the Yomondaka would sing back as the Kakayuni and Pakoa approached the grounds:

> Kakayuni: We're coming to get pigs and money now, give pigs to the mother's side. The father's side won't be the only ones to eat.
> Yomondaka: Our brother's dead, and we're feeling a lot of pain.
> Kakayuni: We felt a lot of pain too, giving birth to this child.
> Yomondaka: Our father walked a long way through the deep forest to come get your mother and marry her.
> Kakayuni: Are you all real men like your father? Will you work a good compensation?
> Yomondaka: You are a lot of men, we are few, pigs and money are small. Some of you will go home disappointed.

As we arrived at the ama, I saw my Tokoyela neighbor Nene singing with the Yomondaka (his wife's clan). His adult son, Yaka, had joined the Kakayuni group, however. Up to this time I had cognitively placed people in one specific clan and had always thought about them in this light. Now the fixed, rigid clan affiliations of my neighbors and informants that I carried around as a mental template were rapidly disintegrating before my eyes.

Over the next several hours the Kakayuni and the Yomondaka argued over who should rightfully claim compensation. Kipan's brother Paul did most of the talking for the Kakayuni. At one point Kipan interrupted to harangue the Yomondaka first in Ipili, then in Enga, with occasional Tok Pisin phrases and a few English "fucks" every now and then accompanied by thrusts of the long spear he held to punctuate his claims (figure 4.5). As the tension built around the proceedings more and more people became agitated, yelling, cursing, and brandishing their axes and machetes across the open space where the pigs were. At this point Kipan, Paul, and Saima pulled out some cash and crossed over to where Angalo was. They handed the cash (about K60) to Angalo to "cool everyone's stomachs" and prevent the outbreak of hostilities. The councilor for Tipinini census ward #2, Isera Pore, who was married into Yomondaka, gave out about K60 to various Kakayuni men to cool their stomachs. Then one of the local schoolteachers, Luke, held up a stick and promised to give a K1,000 pig to the Yomondaka (his paternal clan) if they promised to take care of his mother's people, the Tokoyela—most of whom were claiming their rights to compensation as "Kakayuni" clan members.

Figure 4.5. Kipan (*center, leaning on a spear*) arguing for claims to *laita* pigs.

With these overtures made, Angalo agreed to distribute some pigs to the Kakayuni. Paul received two pigs, and Kipan and Saima each got one pig. The rest of the pigs went to some of the Pakoa who had joined with the Kakayuni and to Angalo's auwene clan, the latter being the people he had originally wanted to give all the pigs to. This example illustrates not only how knowledge of a genealogical system allows one access to different garden lands and social groups but how genealogical knowledge can be used to make claims for wealth during exchange ceremonies. In fact it was Kipan's in-depth knowledge of who was related to the five Kakayuni sisters that allowed him to form a group of men who could disrupt Angalo's plans to give all of his laita pigs to his auwene clan.

Group Connections across Space and Time

As these examples show, there is much flexibility in terms of how people can define themselves in relation to a group. Prior to mining, every clan had abundant land and never enough people; therefore incorporating outsiders was critical for increasing group membership. One of the ways that Porgerans characterize the relationship between land and social groups is with the statement "The land is big, and the groups are small." As such they are intent on recruiting kin members with promises of land so that group size is increased. Large groups have the advantage of

connect to other book

being able to marshal more resources during exchanges and to provide sufficient warriors to either defend the group during tribal fights or, as most people hope, to deter conflict altogether. There are also mechanisms to recruit nonkin, which involve providing land for gardening and houses for varying lengths of time. The goal of bringing nonkin to a group's land is to eventually turn them into kin by marrying them into the group or by marrying their children into the group when they come of age. Nonkin living with a clan are called epo atene in Ipili, which means "those who come and stay." In the eastern Porgera valley in 1999, census data from three hamlets (107 men) that I collected showed that 35.5 percent of males were tenes (agnates) on their own clan's land, 17.8 percent were living as wanainis (descended from a woman), 13.1 percent were wanakalinis (in-married men), and 33.6 percent were epo atenes (outsiders with no kinship link). In sum, just over half of the men in the population were consanguines, fully a third were nonkin, and the rest affines. However, these data do not show that many of these men were pu ipu (coming and going), nor do they capture the temporal variability that would have been dynamic and shifting as some of the agnates would have gone off to live for periods of time as affines with their wife's kin, and some would have lived as epo atenes on the lands of other nonkin friends.

The local conception of social organization and access to land is captured by the phrase *yu towa ate lolane pene* [the woman and her land are mine], which is said by men of an unmarried woman to whom they are not related. Men use this phrase when they are asked if they have a kinship relation to an unmarried woman. A woman is also idiomatically referred to as a "bridge" (*wana ipa toko*; see Strathern 1995) in that she opens up paths between various clan lands, allowing her husband and his kin to garden on her clans' lands. This indicates that wanakalini (women's husbands) and akali wetene (men's wives) are also landowners on their spouse's lands.

One clan alone does not constitute a social group in this area. Rather a set of intermarrying clans with members from all of these groups form a potential set of people who collectively make exchanges, help with gardening, socialize, defend the group territory, and more (cf. Golub 2014). Linguistically it may appear as though it is the activities of one clan that is engaging in these events, such as when people say that Tokoyela fought Undiki, or Tokoyela made a compensation to Pakoa. In reality one group of people from multiple clans is engaging with another, similar group. Biersack (1995b: 237) calls these groups of people "affinal clusters"; they take the name of the clan on whose land they are living.

Table 4.1 Marriage percentages by generation (Tokoyela, Makepo, and Pakoa clans)

	Gen_0	Gen_{-1}	Gen_{-2}	Gen_{-3}
Other Tipinini	42.2	61	61.4	62
Porgera	20.2	8.6	10.5	13.7
Western Enga	20.3	19.9	17.5	6.8
Paiela	0.9	1.9	1.7	0
Kandep/Wage	8.7	4.7	7.0	10.3
Huli	0.9	1.9	1.7	6.9
Other	7.3	1.9	0	0
	n = 109	n = 105	n = 57	n = 29

Examining affinal clusters from the perspective of 342 marriages that occurred in the Makepo, Tokoyela, and Pakoa clans in the eastern Porgera valley (Tipinini area), I found that 49.4 percent occurred with neighboring clans in the immediate (Tipinini) area, 20.8 percent occurred with western Enga, 12.6 percent occurred with other Porgeran clans, 6.4 percent with Enga and mixed Enga-Huli from the Kandep/Wage area, 4.1 percent with western Ipili (Paiela), 3.5 percent with nonlocal groups, and 3.2 percent with Huli from the Tari area. From the perspective of ethnic groups, 66.1 percent occurred with other Ipili, 27.2 percent with Enga, 3.2 percent with Huli, and 3.5 percent fall into the category "Other."

Reanalyzing 300 of these marriages by generation offers another insight into how the history of marriage patterning is shifting in the region. In table 4.1, Gen_0 refers to people who married in the past twenty years or so, Gen_{-1} are their parents, and so on. The most dramatic shift in marriage patterns is the 20 percent drop in marriages within the Tipinini area in the current generation. This drop can be explained by the fact that Tipininians are increasingly marrying Porgerans to their west, while western Enga are marrying into eastern Porgera. In other words, there is a definite reorientation westward toward the mine. The "Other" category shows that other outsiders too are marrying into eastern Porgera in greater numbers. Huli marriages have declined, indicating that many Huli are probably choosing to marry directly into landowning clans rather than the nonlandowning Tipinini clans.

Marriage therefore provides nonlandowner Porgerans a mechanism to try to establish connections with landowning clans affiliated with mining. In most cases women from eastern Porgera are marrying into the landowning clans. There can be intense pressure on young women to stay in these relationships when marital problems arise. Susan Pulaipia was in her early twenties from a family in Kolatika. She had married into one of the landowning clans in early 1998. I often ate dinner at her parents' house and was witness to the unraveling of her marriage and the reverberations it caused among her clan members in Kolatika. Susan described these events to me in September 1999:

> When I first met my husband, I didn't even know him, we'd never even talked, but we met at a *singsing* [traditional dance] in Kairik. He told me he wanted to marry me, and I agreed because he had a job at the mine and was a landowner. When I told Tiwi and Pulaipia [her parents], they were worried as I knew nothing about him. But he gave my clan twenty-eight pigs and K1,000 for my bridewealth [this is double what most Tipinini bridewealth payments are].
>
> After nine months I had a baby, and I wanted to come home to show my family the baby. He wouldn't let me, though. He wouldn't even let me leave the house. He told me I couldn't talk to anyone, even my female in-laws whom we lived next to. He started beating me every time I talked to someone. He kept a stick in the house just to beat me with.
>
> About a week after I gave birth, he came home one morning drunk. I told him to make me breakfast, but he said he didn't want to. Our custom is that women can't cook food until one month after they've given birth. There was no one to cook for me, so I came back to Kolatika. He sent word for me to come back to him, but I told my parents I wanted to leave him and return the bridewealth money. He saw my sister at the Station and told her I was a skinny, dry-faced, *pipia* [trash] woman. He threatened to kill someone from Kolatika if I didn't come back to him. He then told my sister that he would cut me in half if I didn't return. He would keep my lower half and give my upper half back to my clan. He also threatened to rape one of the high school girls from Kolatika unless I returned.
>
> Despite all of this, my clan urged me to return to him. He had given a big bridewealth payment and people were worried about how to pay it back. [Ben and Peter had each received a pig and, as they were now Seventh Day Adventists, couldn't pay back the debt with a pig.] I went back to him, and things were all right for a little while, but then he beat me up again. My clan came to his house. He was raging mad and told everyone to fuck off, I was his wife and he would do

whatever he wanted to me. My clan told me to come with them, but as I started to leave, my husband took our baby from me. Pulaipia told him to leave the baby with me as it was still nursing. He told my dad, "If you want a baby so bad, you should go fuck your daughter!" [The worst insults in Porgeran culture refer to men having sex with their kinswomen.] My husband grabbed a bush knife and tried to chop Pulaipia, but my uncle stopped him and got a deep cut on his hand.

After this my clan agreed that they would return the bridewealth payment. Usually when the bridewealth is returned, they only have to pay half of it. But my clan wanted to keep my baby and get rid of this man permanently, so [they] promised to pay all of it back.

Susan's clan never did return the bridewealth payment. In 2006 she was still living with her husband, highlighting the social pressure to remain in such relationships where the perceived benefits of access to mining wealth outweigh the costs to a woman's health and security, forcing many women to remain in difficult relationships so other clan members can mobilize the potential linkages she offers.

The goals in this chapter were to demonstrate how human labor—work (piape)—is central to the basis of livelihood production and social reproduction in Porgera, and to provide insights into how social groups are formed and connections maintained through participation in compensations and other forms of sociality. At the mundane level of everyday production, work shapes the natural world (the forest) into landscapes that are optimal for human endeavors, and the environment delineates the kinds of human endeavors that in turn shape those environments. Nevertheless human agency alone is not sufficient; ideally Porgeran social-ecological systems depend on the propitiation of spirits. Additionally, while people work to create social groups through marriage and compensations, large-scale ritual activities were also instrumental in tying disparate groups together. These ritual activities had the additional aim of increasing the fecundity of land and society. Their demise has profound consequences for Porgerans' social and ecological systems.

The key aspect of Porgeran social groups is their dynamic nature. There are few (if any) de jure rules that regulate group membership. Group membership is defined de facto through quotidian practices, such as sharing food and labor, living and gardening on a particular group's land, and providing support to that group during warfare, compensation, and bridewealth exchanges. As highlighted in the example of the laita compensation, the channels through which wealth flows are contingent, open to contestation, and dependent upon the ability of a kin group to lay claim to wealth that they rightly believe is theirs. This too has severe implications for the way that mining wealth is distributed in Porgera today. State and corporate processes of "simplification" (Scott 1998) allow mining benefits to circumvent customary processes of wealth distribution.

CHAPTER 5

Spirits: *Yama*

The Cosmology of Water

There is a fundamental principle in Porgeran thought that centers on the states of wetness and dryness. In general wetness and its associated qualities are considered to be a normal or natural state of things, while dryness is a sign of sickness, old age, or infertility. Traditionally, as noted earlier, collective male bachelor fertility cults and individual female maturation practices used water in order to grow young people into marriageable, fertile adults. Water, in both of its manifestations as rain and rivers, is embedded in ideas about spirits, the need for moral behavior in regard to spirits, and moral breaches of reciprocity likely resulting in world-ending calamities. The destructive capacity of water is a common topic due to the numerous rain-caused landslides and torrential rains that swell the waterways and erode stream and river-side gardens. Yet water is also recognized for its creative abilities to heal, promote growth, and combat desiccation and dryness that are symptoms of human illness and ecological decline. Water is also an important metaphor in that its flow is believed to ritually link together diverse cultural groups across the PNG highlands (Goldman and Ballard 1998).

Cosmologically water has the capacity to bestow immortality. Engans (Wiessner and Tumu 1998), Huli (Ballard 1998; Wardlow 2006), and

Porgerans all recite a story about missed immortality when the first baby was born. Piyapa Inguni of Wailya told the story as follows:

> There is a mountain near Mulitaka that has no trees on it. It is only covered with grass. It is called Mt. Mungalo. Long ago, there was a woman on Mt. Mungalo and she had just given birth. Her husband told her that he was going to go get water for the baby from a sacred lake [ipa tawe, "sky water," small high-altitude lakes associated with the sky people]. He told his wife that if the baby cried she couldn't give it breast milk, she had to wait until he returned with the water. While the man was gone, the baby cried and cried and cried. Finally, the woman let the baby drink from her breast. Then from afar the man cried out, "Atambiape [Live?]?" He heard nothing, so he yelled out, "Umambiape [Die]?" The woman responded affirmatively, "Uuuu." When she responded to umambiape, he dumped the sacred water on the ground. Atambiape is the same message that is inside the Bible or that the government tells us. It means we can't fight or get cross. Umambiape has to do with fighting and killing, and we followed this path. If the baby had drunk the water, we would live forever. Since it drank the milk, we are all mortal. (field notes, June 1999)

Water comes from the sky people. Sky women cause rain, and sky men are responsible for thunder and lightning. As a gift of the sky people and as a one-time potential source of immortality, water has the power to wash away sickness and restore health. Water isn't just important for humans; it figures critically in key ideas about Porgeran ecology. A central principle in Porgeran ecological understanding centers around the idea that human and spiritual activities and environmental processes facilitate the movement of the vital life force, called ipane (grease), through people, plants, animals, and the land. Ipane is composed of two morphemes: ipa (water) and ne (the thing that is like X, X being the prefix word before the suffix –ne).[1] In this formulation grease is like water. Just as rivers flow across the landscape and disappear into sinkholes and move through the land to reappear elsewhere, ipane flows through the land into plants, into humans and animals that consume plants, and then is released back into the land when animals are sacrificed and people die. One of the main purposes of ipane is to replenish people, plants, and land that are perceived to be dry. Chris Ballard (2000: 209) argues that the neighboring Huli's concerns with dryness are tied to the specific moisture requirements of the staple that preceded the sweet potato, taro, which is intolerant of dry conditions. This is an apt observation of Porgerans as well.

Human relations with nonhuman spirits are deeply embedded within any comprehensive understanding of social-ecological process. If we want to seriously understand Porgeran practices today in terms of tribal fighting, involvement in mining development, and conversion to Christianity, it is necessary to detail in depth their behaviors regarding the spirit world and its intersections with human endeavors. Throughout this book I have made numerous references to how important spirits are in the Porgeran worldview; in this chapter I deal more specifically with the spirit world, especially in terms of two main classes of spirits, nature spirits and human spirits, for their role in mediating people's relationships with the land and one another.

Throughout the chapters in part II of this book, a cultural emphasis on reciprocity underlies many of the examples. Water and other liquid substances (grease, blood, semen, breast milk) and land are the mediums through which reciprocity among people and between humans and nonhuman spirits is sustained. Failure to respect or maintain reciprocal relations between humans and spirits typically results in negative impacts on production or reproduction. Therefore, similar to what Michael Dove (2011: 169–89) describes for the Kantu' of Borneo, there is a "moral ecology" that permeates principles of reciprocity for Porgerans. Water in its many forms—rain, rivers, ponds—is an elemental force in Porgeran culture and ecology. Water, or ipa in both Ipili and Enga, figures prominently in curing rituals and in ideas about the maintenance of fertility in people, flora and fauna, and even the land. There is an opposition, though, between moving water and still water; water in motion is considered more efficacious in its power to heal and nurture. This concept of motion highlights another key metaphor in Porgeran philosophy that contrasts movement with stasis. In most cases stasis is a result of some form of blockage or stoppage. Most illnesses are perceived as the inability of fluids to move through the body. Cures for sickness either utilize or emulate the free flow of liquids; sitting under waterfalls, drinking large quantities of water, and invoking the names of certain sappy trees in healing spells are just some of the ways that fluidity is deployed in healing. In all of these cases spirits play essential roles and are also linked in prominent ways to water and liquid substances.

Spirits and the Origin of Water

According to oral history, there was once a time when there was no flowing water on the land and the source of water was jealously guarded by a man named Lemeane. The story of Lemeane and his wife, Lakeame, is the most important story that eastern Porgerans tell; they refer to Lemeane and Lakeame as "the first people" who transformed into sky people yet remain the "ancestors" of contemporary eastern Porgerans. The myth about Lemeane is a "charter" myth in the Malinowskian sense in that it encodes social norms and values, but it also highlights the relationship between myth and landscape that Malinowski (1984: 298) discusses in *Argonauts of the Western Pacific*. His concern was with "the influence of myth" on the landscape, how "it colours it, gives it meaning, and transforms it into something live and familiar." In the case of Lemeane, this myth colors the landscape and gives it meaning through his activities that shaped the prominent features of the landscape and the past practices that propitiated him after he became a powerful spirit. Here I provide a compilation of this myth from the numerous versions I heard of it:

> Lemeane and Lakeame lived at a place called Auwalo Anda [about 300 meters north of the main road at the base of Mt. Asienda; map 5.1]. When they were alive, the world was flat, not mountainous like it is today. During this time there were no rivers or springs. Only Lemeane knew where the source of water was, and he kept that knowledge secret. One day Lemeane left to attend a pig feast in Lyeimi [20 or so kilometers north of Porgera]. He told Lakeame to stay home and continue working in the gardens. He filled up several gourds and bamboo tubes of water for her before leaving and told her to not drink the water too quickly. Then Lemeane left.
>
> Lakeame stayed working in the gardens, but the weather turned very hot. The sun was relentless, and within a few days Lakeame had finished all of the water Lemeane had left her. Thirsty, she set out to find out where Lemeane collected water. She followed his footprints to the base of Mt. Asienda and saw nothing except for some vines that men use to climb pandanus trees. The vines lay at the base of a very tall *aiyakena* wild pandanus tree that grew right against the cliff face of Mt. Asienda. Lakeame lashed the vines around her feet and started shimmying up the tree. When she arrived at the top, she could hear a rushing sound but could see nothing except for a flying fox (*kaima*) bone stuck into the face of the cliff. She grabbed the bone and pulled it out, releasing a rushing torrent of water. Within no time at all the water had destroyed all of the gardens around Auwalo Anda and soon started to flood the entire land.
>
> Some hours later Lakeame sat in the top of the aiyakena tree and looked out at the flooded landscape below her. She could see no

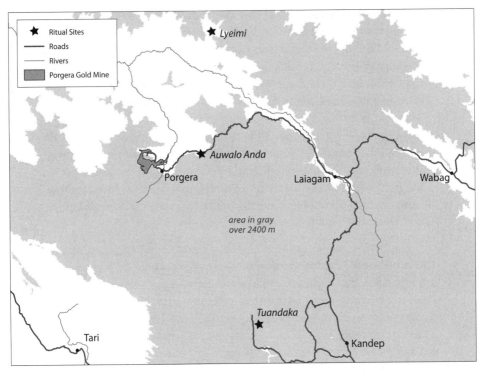

Map 5.1. Ritual sites in the Porgera region.

land anywhere. A *timbo* horsefly flew by her, and she captured it and told it to fly off and find Lemeane and let him know what happened. She tied a piece of her grass skirt to the horsefly so Lemeane would know she sent it. The fly went down to Lyeimi, where Lemeane was at the pig feast. The people were killing the pigs and cutting them up to cook them in earth ovens. The fly flew up to Lemeane and bit him on the leg. As he prepared to smash the fly, he saw the piece of Lakeame's grass skirt. He asked the fly what was happening, and the fly told him that Lakeame had gone to where Lemeane gets water and flooded the land.

Lemeane left the pig feast and since he hadn't had time to share in any meat, the people gave him a large reddish pig named Yia One Panalia to take with him [*yia* is "pig," *one* is "reddish colored," *pana* means "to leave without coming back"; i.e., the reddish pig that leaves without coming back]. He left Lyeimi and started walking home. Night fell and on top of Mt. Kule he lit a torch made out of *daka* (*Piper* sp.). Now daka grows all over Mt. Kule because this is where Lemeane lit his torch. He walked on and saw two mountains, Mt. Kuato and Mt. Tangolo; everything else was under water. He sat on the mountains and told the pig to start digging drains for the water. The pig dug out the Lagaip River and the Porgera River. The water started to drain, so the two walked on and came to Puluma and slept there. At Puluma he dropped his daka leaf torch, and now daka grows all around Puluma. The next morning Lemeane took a digging stick and he and One Yia Panalia dug out the Tupa, Silia, Wateya, and Auwaip rivers [in the eastern Porgera valley]. The water ran down the river courses that Lemeane and the pig had dug and the land started to reappear. At the place where the Auwaip and Wateya rivers come close together, Lemeane tied up the pig [today the site of a huge house-size boulder] and went up to his house at Auwalo Anda.

At this juncture the story takes several divergent twists depending on who is telling it. The full version of the myth can take over an hour to tell, especially if at night around a fire and if a large audience is present. Generally what happens next is that Lakeame runs away in fear of what she has done and journeys to the sky world to live among the sky people. Lemeane, however, falls from a pai oak (*Castanopsis acuminatissima*) tree and dies. His bones turn to stones (like the pig, Yia One Panalia), and two of his cousins from Lyeimi find him and cover him with half-sides of pork to honor him. His descendants are told to sacrifice pigs so that the water will continue to flow off the land and not back up and flood the world. Lemeane is credited with being the regulator of water (Biersack 1998b), and sacrifices were made to him to both "keep him sleeping," for if he awoke he would be upset and cause the world to flood, and "feed him," for during droughts or other socioecological calamities it was believed he was negatively impacting the fertility of the earth and people. In other versions of the story he travels up the *tawe toko* (sky bridge) to live forever with Lakeame, indicating that some Porgerans perceive of them as sky people,[2] which links them to having control over rain.

The last sacrifices to Lemeane were performed in the late 1960s (Gibbs 1977). In the early 1970s Leo Itai, a recently converted Porgeran, destroyed the shrine houses adjacent to Lemeane's bones and had a Catholic priest pour holy water on the site. In 1999 Leo walked me around the site. Overgrown sweet potato vines now covered Lemeane's bones, and Leo had built a small church where the shrine houses had been. A few months after showing me the stones marking Lemeane's body, Leo loaned the government some land adjacent to the main road for the purpose of clearing an area to build a vegetable market. During the land clearing a massive excavator was swallowed in a sinkhole, and only after several weeks was the contractor able to get it out. Walking by the half-buried excavator during this time, Porgerans would repeatedly claim that it was Lemeane that caused it, upset over being awakened by the construction activities. While this was the only contemporary recognition of Lemeane's powers, many elder Porgerans say that there would be no rivers today if it had not been for Lakeame, positing her as a Promethean bringer of flowing water.

Lakeame and Lemeane symbolize an important aspect of Porgeran philosophy in the recognition that it is impossible to separate the biophysical from the metaphysical in terms of how the world works. Lemeane and Lakeame also signify the tension between creation and destruction that is ever present in coupled biophysical and metaphysical realms. Lakeame's gift of freely flowing water was possible only through her failure to listen to her husband (as many people pointed out) and the subsequent destruction of the world. Rituals to Lemeane attempted to balance and regulate his creative and destructive powers of control over hydrological systems. Porgeran ritual efforts involved humans sacrificing pigs to spirits. These efforts were not simple, linear cause-effect attempts at socio-environmental homeostasis (Rappaport 1967) but were full of uncertainty as to how spirits would respond. As a consequence people had to actively read climatic conditions, ecological processes, landscape dynamics, and human biological properties to constantly accommodate uncertain and complex behaviors in spirits that affected the world.

Ritual Sites and Regional Interactions

Auwalo Anda, the ritual site where the control of the land's water balance and the renewal of the earth's fertility was performed, was built where Lemeane's cousins found his body that had turned to stone. Auwalo Anda was one site in what was seen regionally as a number of connected sites in a larger system of earth fertility rituals. The key site in the network was called Tuandaka and was located on the lands of Huli speakers to the southeast of Porgera. Among the Huli this ritual system has been well documented and refers to a set of rituals known as *dindi gamu* (earth magic; see Allen and Frankel 1991b; Ballard 1995: 48–52; Ballard 1998; Biersack 1998b; Frankel 1986: 16–26; Glasse 1995; Goldman 1983: 112–20; Wiessner and Tumu 1998: 186–92). For the Huli the entire ritual sequence of dindi gamu is made up of smaller rituals at sites called *gebeanda* (ancestor house; Ballard 1995: 49) that were located on top of *dindi pongo* (earth knots). These dindi pongo are structurally similar to the Porgeran concept of yu kimbuni (earth joints), in that they are places "where the key strands of the earth's fabric interweave, and so gain strength" (Frankel 1986: 19). People describe the earth joint at Auwalo Anda as a rattan cane with roots that stretch out into the land. The sacrifice of pigs to Lemeane on this earth joint ensured that the grease (ipane) flowed outward along the path of the roots. The Huli believe the earth knots are linked by roots and a thick rattan cane entwined with a python (Goldman 1983: 113), which provides a pathway for the flow of ritual power from one site to the next. These paths are also said to constitute "the physical support of the earth" (Frankel 1986: 19), giving them a function similar to that of the earth joints in Porgeran cosmology.

The Auwalo Anda site in Porgera was also associated with a rattan cane (*kelewa*), and a number of older people who took part in the rituals said that the ritual's purpose was to straighten or clean the roots of the cane so that the land would become fertile again. Similar to the Huli dindi gamu, the Auwalo Anda ritual was performed whenever people felt that fertility in the region, be it of crops, pandanus, pigs, or people, was in decline. The caretakers of Auwalo Anda would send out word to congregate at the ritual site, and people from a number of different clans would bring different items essential for the performance of the ritual. Pore Newe was one of the last caretakers of Auwalo Anda to conduct the ritual. He recalled the site and the performances:

> The people used to have respect for this area [the high-altitude forest, aiyandaka]. People didn't just come up here for any reason. It was dense forest, huge *taiya* trees [*Araucaria cunninghamii*]. No one touched the plants up here. In this area you couldn't kill a mosquito or chase it off. If it was biting you, you just had to let it bite you and fly off.
>
> Underneath the stone that marked Lemeane's body there are the roots [*pini*] of a giant rattan cane [kelewa]. This spot marked the yu kimbuni [earth joints]. When they wanted to clean [*wa pi*] the roots they would clear all the grass away and kill pigs and cook them with vegetables. This would make the roots of the kelewa grow good. The yu kimbuni was the last place that was finished when the ground was created. (field notes, January 1999)

The purpose of the ritual was thus to guarantee that crops, pigs, and people would flourish. Lemeane was also the regulator of rain water. If the rites were performed, then heavy rains would not come and flood everything, nor would long droughts ensue, but enough rain for agricultural purposes would fall.

Gebeanda were sites in the Huli landscape where the cane and python neared the surface of the land and were accessible for ritual manipulation to bring about a time of plenty when people were suffering from drought or disease (Glasse 1995). The gebeanda were surrounded by hoop pines (*Araucaria cunninghamii*) and featured "a cave, a pool or other striking natural formation" (Frankel 1986: 19). At Tuandaka, the gebeanda site closest to Porgera, there is a pool in a cave that signals the level of the earth's fertility. Ritual participants would dip a stick into the pool to see how wet the earth was—"like checking your car's oil," one Huli man from Tuandaka said to me. If the earth was too dry, then the dindi gamu rituals would occur; when one gebeanda site would "see the smoke rising" from another site, the participants knew it was their turn in the cycle to conduct the sacrificial rituals. For the Huli living in Porgera, Tuandaka was not merely a link in a larger dindi gamu network but rather its central point, the crucial lynchpin in the entire universe. "Tuandaka is the center of the world, it's where all the dindi pongo come together," another Huli man related. Another said, "PNG is at the middle of the world, all the rest of the countries are around the sides. Tuandaka is in the center, the middle of the middle. Inside the cave, when they went to kill pigs they would retie the four rattan canes that hold together the four corners of the world."

The python wrapped in cane is called Puya Dama (Python Spirit; *dama* in Huli is equivalent to yama in Ipili) by Huli speakers; many of them mentioned different branches of the python going off to places where today there are major resource development projects, such as the goldfields at Porgera and Mt. Kare, the oil and gas fields around Lake Kutubu, the gas fields at Hides near Tari, and Mt. Hagen, the economic center of the highlands. The python's head was located at Tuandaka, and during the ritual to renew the earth's fertility, massive amounts of food were fed into a hole in the limestone inside the cave that was said to be its mouth. In addition to the python in the cave, there were two women who were the caretakers of the dindi pongo.

The story of where the two women in the cave came from is connected to the story of Lemeane and Lakeame. As I mentioned, in some versions of this story Lakeame travels to the sky world, becoming a sky woman. In other versions of the story, Lakeame (who is either Lemeane's sister or second wife) and Ipali (Lemeane's first wife) travel from Auwalo Anda (Lemeane's house) to Tuandaka. The following story was told to me by three elder men in 1999, Pakena Ipara, Alua Maleya, and Kuraia Yakob. The first two men's mothers' clans were from the Tuandaka area:

Pakena: The two women [Lakeame and Ipali] left Auwalo Anda by themselves. The two came upon a place called Ingau. They followed the Pongema River upstream from near Suyan. Lakeame was using a walking stick made of *kenda* wood and built a bridge over the river with it. Today this bridge is a stone that can be seen there. They crossed the river and there were taiya trees [*Araucaria cunninghamii*, hoop pine] and *aiyake* trees [*Ptychococcus lepidotus*, black palm] on a mountain there. Lakeame went and sat down among these trees. The two tried out this area, but they thought that this place wasn't comfortable enough so they continued on. When Lakeame left Auwalo Anda, she was carrying seeds of aiyakena [*Pandanus* sp.] in her net bag, she was carrying *yupi* [*Areca macrocalyx*, wild betel nut] seeds and *pima* seeds [*Pandanus antaresensis*] too. All of these grow now at Suyan where she sat down.

The two came up to Pakapuwa. They followed a river and came up to Mt. Maipia. Then they went down to Ipakola. At this place is a mountain called Lokaipiliwi. *Maipia panga tuku wito lolopita* is an expression in Huli that means "I am saying that the door at Maipia will open." Lakeame was carrying one pig with her as well. She had a *paiyatake* stick [a paddle-shaped stick], and with this stick she beat the pig until it was dead. While she beat the pig she said the Huli phrase that made a door appear in the side of the mountain.

She left this area and came up next to the Wage River. She removed a cassowary quill from her nose and built a bridge across the Wage River. There is a stone here today where this bridge was. Next she came up to Pangalini; from here she went to Tuandaka. Tuandaka is at a place called Yuama. People used to kill pigs and carry them inside this mountain at Tuandaka. At Tuandaka she searched around for a good place to sit down. She thought that Tuandaka looked like a good place to sit down, and she thought she could live here. She wanted to find the best place to live, though, so she went down the Andapali River [near Margarima]. She sat down here at Paiapaia. At Paiapaia she came back again.

Alua: Where she sat at Paiapaia, there is a mountain called Yambiparo. There is a stone on Yambiparo that indicates she sat down here, her back was facing toward Margarima. Before, a smoke would come up from Paiapaia. This smoke marks a gas that comes up here.

Ben, my research assistant: Today people light this gas with matches when they walk by it.

Pakena: There is a place called Gu Ipa [literally "smells bad water" in Huli; because of the gas]. This is close to the Wage River. Ipali sat down here. At this place there is diesel [*disilini*] in the water.

Kuraia: When they killed pigs here . . .

Pakena: They said when they killed pigs here, smoke would come up at Auwalo Anda. When we killed pigs here [Auwalo Anda], smoke would come up at Tuandaka. The head is Tuandaka, the tail is Lyeimi, the middle is Auwalo Anda.

In this story Pakena, Alua, and Kuraia detail the two women's perambulations from Auwalo Anda in the eastern Porgera valley to Suyan (another ritual *yu kimbuni* site, today where the expatriate and non-local mine workers are housed), to Tuandaka, to the oil and gas fields of Hela Province where the multibillion-dollar liquid natural gas project is currently under development, and to Paiapaia Tepetepe, another important *gebeanda* site. Contemporary features of the landscape, such as dominant tree species and large rock outcroppings, provide evidence of the women's journey. The final comment by Pakena highlights the notion that a python was central to the linkages between the ritual sites at Auwalo Anda and Tuandaka. Porgerans also note that Auwalo Anda was linked to another ritual site among the western Enga at Lyeimi (where Lemeane went for the pig feast). I was never able to determine what occurred at Lyeimi, although it was the home of Wambilipa, an important prophet, who was responsible for the post–World War II cults discussed in chapter 2.

It is not known whether Huli from Tuandaka in the pre-mining era would have accepted that the two women in the cave were from Porgera. Given today's current resource development projects, it certainly aids in creating links between Huli and Ipili speakers through a shared ritual and mythical network (cf. Biersack 1999b). Similarly for Porgerans, based on some elders' stories, the connections between Auwalo Anda and the Huli earth magic system seem to be more recent. However, just as there is no harm in Porgeran stories linking Lakeame and Ipali with the two women in the cave at Tuandaka, there is no harm in Huli tying Auwalo Anda into a powerful, regionwide ritual network from Porgerans' perspective.

As noted, the python at Tuandaka is marked by mountains that cover its several bodily branches. A few months after Pakena told of the two women at Tuandaka, his son, Friday, was talking about the gold at Mt. Kare and Mt. Watukati (the source of Porgera's gold): "Wherever those two women went there's gold now. Tuandaka was the head, Mt. Kare was the stomach, and Mt. Watukati was the tail." While there currently is no gold at Tuandaka, people expect that it will be discovered soon, for gold, silver, oil, and gas are said to come from the shed skin, urine, and feces of the python. The Huli at Tuandaka cite an origin story (malu) in which the kindness of a young man to an old woman resulted in mineral deposits:

Agnes Bebi: Ten young men were living together. Nine stayed home, and one left to go hunt marsupials. While hunting, he saw a tree with a giant hole that was all scratched up around the edges. The man climbed up and stuck his head in the hole in the tree. Something grabbed his head and pulled his body into the tree. He didn't return, and three days went by. Then another brother left and followed his first brother's footprints. He was looking for marsupials and saw the same hole. He climbed up the tree and stuck his head in the hole and something pulled him into the hole by the head. Then another brother left. The same thing happened to him. Then another brother, and another, and another, et cetera.

Finally one brother was left. He was thinking of his brothers and wondered where they all had gone. The brothers had a huge pig that the youngest killed and cooked. He carried the cooked pig and set off, following his other brothers' footprints. He walked and walked and saw an old woman sitting beside the road. She lived in a rotten old house. She had no firewood; she had to use her own dried feces to build a fire. The old woman said, "No men ever come here. What are you doing here?" The boy gave her some pig meat from the choicest part of the pig. He built her a new house. He cut firewood. She said, "No one ever helps me. What can I do for you?" The boy said, "Nine of my brothers left and never returned. I'm looking for them."

The woman told him to go cut a tree and make a pair of fire tongs. She told him to heat a round stone in the fire. He heated the stone until it was red hot. He took pig grease and wrapped it around the stone. The woman told him to carry the stone to the tree with the fire tongs. She told him to climb the tree. She told him that something would try and eat him; he should drop the stone into its mouth when it tried.

The man did as the old woman told him. The thing opened its mouth and the man dropped the hot stone into its mouth. The stone was cooking the beast. The tree was shaking and shaking. The tree broke apart and a giant python came out. The python had eaten his nine brothers. It was crawling and crawling all over—it crawled to Tuandaka and went inside the cave. The python's trail is now the Tase River. The Wage River flows on top and the Tase River goes underground at Tuandaka and goes underneath the mountains and the Wage River and comes up at Tari. Paiapaia Tepetepe is the place where the water shoots up. There's a gebeanda at Paiapaia Tepetepe. When we sacrifice pigs at Tuandaka, smoke will come up at Paiapaia Tepetepe and they know we are holding the dindi gamu.

Pokali Kelu: The river is like a pipe. When we sacrificed pigs and gave them to the python in the cave, smoke would come up at Paiapaia Tepetepe. When the Huli at Tari sacrificed at Paiapaia smoke came up at Tuandaka.

Muli Kili: All the oil is the urine of the python and the gold is his shit. (field notes, October 1999)

In these accounts contemporary mineral deposits are the result of human interaction with a python spirit. The kindness of a young man to a poor old woman removed a dangerous creature that had consumed the young man's entire family. Once the python moved from the tree to the ground it shifted from the consumption of humans to the consumption of pigs and products of the garden. A mutual dependence developed: the python relied on humans for food, while humans needed the python to renew the fertility of the land. As Muli Kili said, "We killed pigs and fed the python so that it would reciprocate and put grease back inside the ground." Some Porgerans refer to the python at Tuandaka as the Diamond Snake, the source of all the mineral wealth in the region. Exploratory activities by mining companies are considered attempts to find the python's head, the "real" source of mineral wealth, which would obviate the need for mines and wells as the python would produce unlimited wealth for its owner in the form of gold, oil, and even diamonds.

In a certain sense, mineral and petroleum wealth is understood to be a reciprocal payment from spirits to humans. The dominance of transnational mining and oil companies is a source of concern, although this fact too was foretold. At Tuandaka the progenitor of the Huli people, a man named Hela, had four sons: Hela Huli (the Huli people), Hela Opene (the Enga and Ipili), Hela Tukupa (groups to the south of the Huli), and Hela Helope. "Hela Helope are the white men," said Agnes. "Hela Helope left and took half of the old stories with him. The other three sons were left with the other half. Our fathers told us that one day Helope would come back for the gold at Tuandaka. We couldn't deny it to him, it belongs to us all." One man from Tuandaka worried because "all of the mountains around here are places where the python lives. Under each mountain is a part of its body. If the gold is removed the python will become angry and twist and thrash and all of the mountains will break and fire will burst out." The fire would not be confined to this region of the PNG highlands, he stressed, but would culminate in the destruction of the entire world.

Kupiane, the Python at Porgera

The logo of the Kupiane Yuu Anduane Pty. Ltd. (the Porgera Landowners Association) is a giant red snake coiled several times around a mountain range with a pandanus tree, which is the logo of the Porgera Gold Mine, reflecting the mine's location in the upper rain forest adjacent to Mt. Watukati (figure 5.1). The snake is Kupiane, the python spirit that lived in the heart of Mt. Watukati and to which groups in the western Porgera valley used to make sacrifices. Today Kupiane is said to be the source of Porgera's gold (Golub 2006) and is linked to the Tiyini, one of the principal landowning clans associated with the Porgera Mine. In the eastern Porgera valley around the Auwalo Anda site, people are aware of the Kupiane story. The stories that eastern Porgerans tell about the origin of gold are quite different from the version told by the Tiyini. The differences between the two stories reflect some of the tensions over resource development at play in the valley today. In what follows, I briefly present the Tiyini version of the Kupiane story based on several versions I heard, supplemented by the anthropologist Alex Golub's (2006: 247–49) version, drawn from his fieldwork among Porgerans around the mine:

> One day Tiyini and his wife left their house to attend a pig distribution at Suyan, leaving their two small children behind in the house. In the evening a heavy rain developed and flooded the rivers preventing the parents from returning home. From his vantage spot on the top of Mt. Watukati, Kupiane [in his human form] could see the two children sleeping alone in the house and given the bad weather decided to go look after them by building a fire and cooking them food. The parents, meanwhile, were sleeping outside and thinking about their children and were jealous of how warm and cozy they must be. The souls [tandini] of the parents left their sleeping bodies and went to attack the children from envy. [Ipili believe that the spiritual form of the human body, the tandini, leaves the body at night during dreams.] When the parents' spirits arrived at the house, Kupiane turned into a giant python and scared the spirits away. The next morning he told

Figure 5.1. The logo of the Kupiane Landowners Association, with Kupiane encircling the logo of Porgera Joint Venture.

the children to tell their parents that Kupiane protected them through the night and that their parents must sacrifice pigs to him. He told them that when a particular yakati tree [*Ficus dammaropsis*, Highlands breadfruit] on Mt. Watukati blooms, then the people should sacrifice pigs to him. He then turned into a python and crawled into the hole in the mountain where the pig meat should be offered.

For generations the Tiyini sacrificed pigs to Kupiane. Then in the mid-twentieth century, a Tiyini man, Kipu, and two of his friends, Pingipe and Esape, started to make prophecies about the future (Golub 2006: 253). Pingipe told everyone that at his place on Mt. Wangima (adjacent to Mt. Watukati) a tree would grow that birds from all over the world would come to eat at. When mining development started some decades later, Porgerans reinterpreted Pingipe's prophecy to mean that what he had foreseen was the coming of gold mining. Once the source of the gold was tied to the very mountain where Kupiane lived, it was understood that Kupiane had communicated this message to Pingipe. Later, in the 1970s, Kipu's son, Ambi, sacrificed pigs to Kupiane at the time that the first mining exploration started. According to Golub, Kupiane appeared in a vision to Ambi, telling him, "You eat" (259). When the mine opened in 1990, Kupiane appeared again and told him to host a feast in celebration of the beginning of gold mining "and that the spirits inside the mountain would celebrate along with him" (259).

In the eastern Porgera valley around Tipinini, the story of Porgera's gold starts with two sisters who live in Umambo, a hamlet in Tipinini. In December 1998 Tisapa Yawe related this version:

> One day the older sister went to gather some takae plants [Oenanthe javanica, an edible parsley-like herb] along the Porgera River. While she was gathering the plants, she saw some pig intestines floating in the water. She cleaned them and put them in her net bag. She wondered why someone would throw away good pig intestines and started following the stream upriver to see who lived there. She followed it all the way to Mt. Watukati.
>
> The sister came upon a huge cliff with a rope hanging down from it. She climbed up the rope and saw a house up there on top of Mt. Watukati. There was a pig cooking in a mumu but there was no one around. Then a huge man came walking up [Kupiane]. The woman was sitting at the base of some cane grass. The man saw the woman and came over to her. He asked, "How did you get up here? There's no road." The woman said she had climbed up a rope, and then the man said that he had a pig cooking and that they would share it. He took the pig out of the mumu and cut it in half. He told the woman not to eat any here. He told her to carry it home and eat it there. She carried the half pig and the stomach and intestines home. [Men don't eat pig stomach or intestines.]
>
> When she got there her younger sister asked, "Hey, who gave you that pig meat?" But the older sister didn't tell her. Every day the older sister would go do this, but the younger sister refused to eat any of the pig. The younger sister said she wanted to go get the pig herself and then she would eat some. The older sister said that she alone was the one who could go get pig meat. The younger sister pleaded with her, so the elder said, "Okay, you can go. But you must do as I tell you." She told her to go to the stream and get the pig stomach and clean it. Then she told her to climb the rope and go sit near the cane grass, and then the man would mumu a pig and give her half. He would tell her not to eat any and to take it home.
>
> The next day the younger sister left, found the pig intestines, cleaned them, and then climbed the rope as her sister had instructed. She didn't see anyone around, but there was a pig cooking in a mumu. She cooked a little bit of the pig intestine for herself and then sat down in the cane grass. The man came up to her and said, "Hey, who

are you? I think you're another woman. Why did you eat here? You broke the law of this place." The man went to the pig and took it out of the mumu. He cut it in half, and as he was doing so the sister ate some of her half. Then she returned home. The older sister asked her how it went, and when she found out what the younger sister had done she was very angry.

The next day the older sister herself went. When she got to the stream there were no pig intestines there. She went up to where the rope was, but the man had cut it. She searched around for a path up Mt. Watukati and eventually found a way to climb up. When she got there, the man wasn't there. She saw the footprints of a man and pigs heading off toward Mt. Pangalini, and then to Mt. Kare. Today, where the man used to throw the pig intestines is the place where they found gold.

Several months later Tisapa and I were talking about this story again. He said that the sisters, Lipime and Napime, were actually the daughters of a python spirit named Taiyundika, the python spirit associated with the gold at Mt. Kare (see also Biersack 1995b, 1999b; Clark 1993; Wardlow 2001).[3] According to Tisapa, Taiyundika kept destroying his two daughters' house by blowing the roof off. So one day they hid near the cave where he lived and when he put his head out to blow off their roof, they cut off his "lips" with a bamboo knife. The python thrashed around, fell into the Porgera River, and then crawled back into his cave and went underground, where his shit produced the gold at Mt. Kare. Tisapa ended this story by saying, "This is why we always say that women are the source of gold."

Eastern Porgerans, like the Huli to their south (see Clark 1993; Wardlow 2001), are largely left out of the benefit streams that come from gold mining in Porgera. The way that eastern Porgerans account for the gold deposits in Mt. Watukati and Mt. Kare is to link two ancestral sisters to the creation of gold. They also depict Kupiane as generous, giving half of a pig to the eastern Porgeran women. Eastern Porgerans are, of course, fully aware of the origin of gold stories told by landowning clans around the mine. Their own version of events is a "hidden transcript" (Scott 1985) that allows them to mythically connect themselves to a contemporary source of mineral wealth. (In an attempt to bring even more people into the ambit of mining development, on one trip to Mt. Hagen, the driver—a Hagen man—told me, "This snake, Kupiane, his body starts in Porgera, goes through Southern Highlands [Province; i.e., Tuandaka], and ends in Mt. Hagen.")[4] The eastern Porgera Kupiane story also reflects the practices of social organization pursued by eastern Porgerans who attempt to marry their sons and daughters into landowning clans affiliated with the mine in order to share in the mining wealth. As such, in both a mythical and pragmatic sense, women (as daughters, sisters, and in-laws) are indeed the source of gold for eastern Porgerans.

Spirits, Souls, and Intentions

While the yama spirits shape Porgeran ideas about the landscape, regional social and ritual relations, and contemporary resource developments, there is a whole other class of what are also called yama in Ipili but pertain to the Western concept of souls. These spirits, what many Christianized Porgerans call their souls, form a broader category than the Western, Christian concept and are called tandini in Ipili. To better understand the cosmological principles that guide human and spirit interactions, in this section I highlight the interplay between biology and metaphysics in Porgera. Physiologically the most important organ in the Porgeran body is the *yamapane*, which includes the heart and lungs. The yamapane is also the seat of thought or reason (*nembo*) and emotions (*mini*). Generally thoughts and emotions stay within a person's body (*angini*) and can be known by others only through a person's actions and words. However, emotions and thoughts can also leave one's body, at which time they generally cause harm. More specifically it is the thoughts and emotions related to greed and envy that result in spirit attacks. In one form of spirit attack, called *yama au* (which roughly translates as "spirit carried"), the tandini leaves a person's body and attacks someone it is envious or jealous of. Yama au attacks most often stem from one person thinking about another person eating a particular coveted food item, typically meat or nut pandanus. However, in November and December 1999, yama au court cases were becoming more common as the heavy rains of that year reduced sweet potato numbers. Peter's wife, Tinka, sued a neighbor after Tinka dreamed that the neighbor was fighting her over a cooked sweet potato that Tinka was preparing to eat. Tinka woke up screaming and realized that her neighbor had sent yama au. While Tinka was able to fight off the attack, she claimed that their son, Wakaini, got sick and remained so for three days while his body was trying to fight the yama au the neighbor had sent. She later said, "Can you believe that? She sent yama au over a sweet potato."

Another symptom of greed or covetousness is when a person's own body becomes ill because of constant longing for meat, money, or even love. This illness, called *uyenda*, causes the inside of a person's body to "dry out." Under normal health conditions, a person's yamapane (heart and lungs) is suspended in a watery matrix. However, due to intense desire the yamapane shrivels and the watery matrix dries out. Left untreated, uyenda can kill a person. Cures require ritual intervention and the consumption of pig grease, traditionally; today ritual experts often use tinned fish. (Lightning is said to be evidence that sky men—tawe akali—are fighting uyenda as it also has a black, shiny spirit form when it is not in a person's body.)

Another way that thoughts can leave a person's body is when the tandini leaves after death or during sleep. One morning over coffee with Peter, we were discussing why Porgerans sleep fully clothed while white people don't. Peter was disdainful of my habit of wearing only boxers to bed and pointed out to me that while we're dreaming our tandinis are out of our bodies and traveling around. He said he wouldn't be able to tolerate the shame (yala) of being seen running around in his underwear by other people's tandinis in the dream world. The Porgeran concept of a tandini is different from what the English word *spirit* or *soul* conveys. A tandini is a person's spiritual body. It is the material form in which spirits can interact with a person. It is also the thing that allows another person's negative emotions and thoughts to cause a person to become sick or die. In this case, tandini can turn into yama spirits, especially when a man's and woman's tandinis combine in the night and turn into a deadly *wanda yama* (woman spirit; see *Wanda Yama*).

The heart and lungs are examined in postmortem autopsies to determine if any particular clan is responsible for killing the deceased. Often, after dying, men are cut open and ritual experts (akali andi anda, "a man who sees/knows") examine their lungs for pus (lingi). If pus is found, then divination rituals are used to determine which clan was responsible for killing the man. If no pus is found, the person is considered to have died of natural causes. Frequently the accusations of intentional murder come years and even decades after a person was injured in a tribal fight. In one autopsy performed on an elderly Tokoyela man in 1999, the autopsy experts determined that he had died from an arrow wound he received in the 1970s. As they described it, a small piece of arrow had broken off in his body and over the next couple of decades traveled through his bloodstream, eventually becoming lodged in his yamapane and killing him. When the Tokoyela demanded compensation from the clan that had shot him with the arrow, the accused clan acknowledged their culpability in his death and made a payment of pigs and money to the Tokoyela. This is fairly common, as to not make compensation would renew hostilities between the two groups.

Today chest X-rays are commonly used to find pus spots in people's lungs. I witnessed several domestic violence court cases in which women brought chest X-rays to village court meetings to have the autopsy experts look for spots that may have been caused by their husbands beating them. Not everyone agrees with this procedure; as at one such trial Ben whispered disdainfully in my ear, "Ol lapun i no save long X-rays na ol wokim giaman toktok long ol dispela samting" (The old men don't know anything about X-rays and are just bullshitting about all of this.) The word yamapane also appears in many words having to do with positive and negative emotions. "To despair" is *yamapane apia* (literally, to throw away one's yamapane). "To be happy" is *yamapane epele a* (to like one's yamapane). "To be sad" is *yamapane kenda a* (to have a heavy yamapane). "To be brave" is *yamapane tu pi* (to have a stubborn yamapane). A good friend can be called a *yamapane yene* (he or she has/makes yamapane).

One afternoon I was sitting with Ben and his father, Pes, outside of Ben's house and we were discussing what kinds of things had nembo (thought, will, consciousness) and tandinis. I was asking them about dogs, birds, trees, rocks. I learned that it is impossible to have nembo or a tandini without having a yamapane. Only creatures that have a heart and lungs can think. Both dogs and trees are alive, yet only dogs have nembo because they take action in the world. As Ben said, "Trees don't make decisions. Dogs decide where to walk, what to eat, when to bark, and so forth." To have a tandini and nembo means that a living entity makes its intentions manifest, but nembo also has an element of moral and proper intentionality to it. At about this time June, a young, fairly severely mentally disabled woman, and her mother, walked past us along the path to their home. Ben commented after she had passed by, "Nembo nateya" (She doesn't have nembo). I argued that of course she had nembo, that while she couldn't talk, she certainly made decisions. Pes disagreed with me and said that she had a tandini, but like a yama spirit, no nembo. According to Biersack's (1991: 234) work among the Paiela valley Ipili, nembo "chooses and desires, 'sorting' the 'good' from the 'bad,' selecting between alternative courses of action and their projected outcomes. In the absence of rational (and in the most important cases moral) decision making, behavior and its effects become 'random' (*ambe pene*)." Rational decision makers have minds that are "straight" and "good," and, importantly, strong passions can rob a person of nembo.

Spirits too can confuse people's thoughts or make them act inappropriately. The most common way this is expressed in Tok Pisin is *tambaran pulim mi* (a spirit pulled me). This same phrase is also used to describe spirit attacks. Occasionally people use the verb *to hold* to describe attacks by spirit. What is being held or pulled by the spirit is the person's tandini. Lata Peapen, Peter's cousin and clan brother, witnessed a wanda yama attack in the early 1980s. In this case three clan brothers, Tisa, Elera, and Wape, went hunting in the upper rain forest for marsupials to present to Elera's new in-laws. They spotted a *wapisia* (Long-beaked Echidna, *Zaglossus bruijnii*), and the dogs started chasing it. It climbed up a cliff, and while climbing after it, Tisa slipped and fell down the cliff, badly injuring himself. Elera and Wape panicked, assuming that a wanda yama had attacked Tisa, and went to Elera's in-laws' home, leaving Tisa in the forest. They told everyone Tisa had died, and when the word got back to his hamlet the village councilor gathered a group of men, including Lata, to bring his body home. When they arrived at the base of the cliff, Tisa was no longer there. They searched around and eventually found him in a tree nearby, barely alive. They knew that a wanda yama had attacked him and put him in a tree. They made a litter and started carrying him back to the hamlet, but night fell before they arrived. They built a bush house in the forest and went to sleep. Lata remembered, "In the middle of the night the wanda yama came to take him away. She was pulling on his legs and we were pulling on his body. We couldn't see the yama, it was like an invisible magnetic force [*pawa olsem magnet*]. We were close to ripping Tisa in half."

Eventually the pulling stopped, and when they got Tisa to the hamlet in the morning they decided to perform *patali* on him. Patali is an indigenous medical procedure to remove pus (lingi) from the yamapane that the doctors at Paiam Hospital disparagingly refer to as "bush surgery." An incision is made along the patient's ribcage; the membrane (pleura) is punctured; and bird feathers are poked into the pleural cavity to remove the pus from the yamapane. A leaf bundle keeps the wound open for several days, until the bird feathers come out clean. Lata said, "Just as I was pinching his skin to make the cut, his body started jerking because the wanda yama had gone into him. He jerked a few times and then died right there." While Elera and Wape eventually compensated Tisa's kin for leaving him alive and hurt in the forest, people didn't blame them for their actions as their thoughts (nembo) had been disrupted by the spirit attack on Tisa. Prior to the patali procedure, *apano kamo* (a protective spell and divination technique; see *Wanda Yama*) was performed to determine who sent the wanda yama. Tisa's mother, Piton, and his sister's husband, Maso, were found guilty. Elera and Wape demanded that Maso recompensate them for the pigs and money they gave to Tisa's kin (who were actually Piton, Maso, and others!). Maso refused, so Elera and Wape broke into his trade store in the night and stole several hundred dollars' worth of supplies. Fighting between the two subclans lasted three weeks, and two warriors on each side were killed. After a time compensations were paid and peace returned, but as one of Biersack's (1991: 256) Paiela informants remarked, "When we war, we imitate the *yamas*." Biersack notes, "*Yamas* are the corruptors of human will, cause of the evil they symbolize" (256).

Wanda Yama

An interview with Kipan Wuambo in January 2000:

Jerry: What is *wanda yama* [woman spirit]?

Kipan: When a man has some pig or marsupial meat, and a man or woman thinks about that meat, their thoughts will go attack that person and we call that wanda yama.

Jerry: Why is it a woman? Why not just a yama?

Kipan: Because men don't usually share meat with women, especially when they're hunting. Some men share meat with their wives and other women; these men will be all right. The thoughts that people send out about meat and other things they crave are yama au. If two women send yama au, a man can get rid of the wanda yama by using apano kamo. If it's two men that send it, a man can also use the kamo. But if a man and a woman together send yama au, this person will die right away—there's no hope for them. Also, wanda yama can't kill a man in the lower rain forest, but if a man goes to the upper rain forest and a wanda yama attacks, he'll die. The reason that wanda yama can't kill a man in the lower rain forest is because of the yu angini wanda [earth mother woman]. She will kill the man instead. When one man goes to the forest alone, he won't get attacked. If two men go, one will die and one will come home. If a man and his dog go, the dog will die and the man will come home. Pigs, dogs, children, and women can all be killed by wanda yama. None of these has the yama au sent to them; instead the yama au is transferred to a man's pig, or dog, or children, or wife, and then they will die.

An interview with Pes Bope in July 1999:

Pes: When men go to the forest for marsupials or pandanus nuts, they always work apano kamo [*apano* leaf magic]. A ways up the trail they'll roll up a bundle of leaves and say magic spells over it and tie it up. Then they place the roll on the trail and cut it into three equal sections. The left-hand side they hide to the left of the trail, the right-hand side to the right. The middle they carry with them and make a friction fire on top of it to start the fire at their bush house. When yama au comes along the trail, it will stop at the place where the two leaves are. It will wait there because it will think that this is where the hunters are sleeping, but actually they are way off somewhere else. This is how we trick yama au. Another way we trick yama au is we take a huge bundle of ferns and hang it in a tree. In the kamo we say the bundle is a marsupial and tell the yama au to come here and wait for us to come back and cook the "marsupial" with them. The yama au will stay here instead of attacking us in the forest.

Jerry: Do you come back along a different trail? I would think that the yama au would be mad at you for tricking it.

Pes: No, we come on the same trail. Remember, yama au is just thoughts of men and women [nembo]. It will only attack you if it comes to you when you are in the process of trying to get marsupials or pandanus or cook them. When you come back, you've already hunted or gathered so the yama au will just go away. You won't get hurt, or not find any marsupials, or your dog won't go wild, those kinds of things.

Spirits and "Eating"

In several publications Biersack (1995a, 1998b, 2005, 2011b) has argued that for the Ipili human life is governed by a "sacrificial principle." What she means by this is that the maintenance and reproduction of human life depletes a person's life force. Gardening, hunting, childbirth, economic exchange, rearing children, and so forth exhaust a person to the point that death is inevitable. While most of us are aware of this at a subconscious level, for Ipili speakers it is a central preoccupation. Children are said to literally "take the place" (*panda mia*) of their parents (Biersack 2011b: 230). The principle of a life for a life also is central to rituals designed for healing, in which pigs are killed for sick people who are being "eaten" (*na*) by spirits.

The signs that someone has spent his or her life in toil under the sacrificial principle are manifested in that person's outward appearance. Wrinkles, white hair, and loss of vitality are all indicators of ageing but are also indicators that one has depleted one's ipane through the work of production and reproduction. By producing children, parents gradually diminish their own life force. Agricultural activities also involve the flow and transfer of grease. All land contains varying amounts of yu ipane (land grease). Over time, as land is cultivated, a garden eventually loses its grease, so that a new garden must be planted where there is more grease. Over even longer periods of time there is a general consensus that the grease has been exhausted from the entire region, and varying intensities of intra- and intergroup rituals are required to revitalize the world.

Every clan has a sacred pool in the high-altitude rain forest above the hamlets where the spirits of deceased ancestors go to live after death. The pools are called *ipa ne*, literally "the water that eats."[5] Traditionally illnesses were (and sometimes still are) believed to be caused by deceased clan ancestors. Various rituals were performed to feed the ancestors, either by offering them "food" in the form of aromas of cooked meat, the blood of pigs poured into the ground, or the actual delivery of meat to the ipa ne. In the case of the latter, multiple informants describe the water in the pool rising up to snatch the bundle of meat from a long pole that ritual supplicants held over the water. Today attitudes about these pools are varied. In chapter 4, I described a long-abandoned ceremony called a kolo that was held in 1999 by the Paiam clan on the site of the Paiam Supermarket (see Jacka 2001). The building of the supermarket had destroyed the Paiam clan's sacred pool, and when food and drinks mysteriously disappeared from the supermarket the Paiam held a kolo to placate their angered ancestors. After a group of young men took me to see the Pakoa clan's sacred pool, I awoke the following day with my first migraine headache, which everyone attributed to our having disturbed the clan's ancestors. A few months later I accompanied Ben and Pes to the site of the Tokoyela's sacred pool. It had long since silted over, and Pes nonchalantly walked all over the grassy surface of the former pool, chopping at the grass with his machete while telling me how afraid they used to be of this area due to the ancestral spirits.

Generally the tandinis leave the living alone, although occasionally the spirits of parents might attack their descendants (children or grandchildren), making them sick, especially if a person had slighted one of their parents shortly before he or she died. The first ethnographer to work in western and central Enga, Mervyn Meggitt (1956: 114–15), wrote the following about spirits of the recently dead and the clan ancestors: "Like ghosts of the recently dead, clan ancestors are generally regarded as being neutrally disposed towards their descendants until the latter antagonize them. Then they become downright malicious, punishing the clan with crop failures (e.g., of sweet potatoes or pandanus), high mortality among children and pigs, and/or defeats and deaths in fights. I gather they are never benevolent or positively helpful, nor do there seem to be strong moral overtones in the punishments." Much pre-Christian ritual was oriented toward appeasing these spirits through pig sacrifice. Pes described these rituals to me:

> We would build a small house in the forest near the ipa ne and kill a pig there and cook small pieces of fat and meat in the fire. We would call the names of recently deceased ancestors and say, "I'm giving this pig to you." We would call out the names of many ancestors and cook small pieces of fat or meat from the shoulder [maputane]. If a female was sick we didn't cook maputane, it was just for males. When we cooked the pig meat, we'd say, "You can't let my child get sick again, you must look out for them." While cooking the meat, we would also say things that would warn us of trouble, like "You must make the sun shine while the rain falls [a sign that an enemy is approaching or something bad will soon happen]. You must tell me when an enemy is coming by making a bird cry, or if you make a tree break and I hear it, then I'll know. If you do these things then I'll know an enemy is coming." (field notes, July 1999)

With conversion to Christianity, most Porgerans are less concerned that the tandinis of their ancestors are attacking them. When people are killed in warfare, though, it is a different matter.

The tandinis of men killed by enemies do not stay peacefully in the ground; instead they visit the dreams of their brothers, real and classificatory, although they are unable to communicate their intentions. When a man has dreamed that a tandini is trying to communicate with him, he and his clan brothers go late at night to the site near where the man was killed. They place on the ground a bamboo tube filled with pig fat and then call out for the tandini to send them a sign that it is in the area. If an ant or termite approaches the bamboo tube, then the men know that the tandini wants to communicate with them. Later the men who contacted the tandini go to the house of a kinswoman of the deceased man, who is called an eka aiyapu (literally "bird prophecy"), or a medium for the dead (cf. Gibbs 1975: 35). While sitting around the fire in an eka aiyapu's house the woman calls out for the tandini of her deceased kinsman while burning small pieces of pig fat in the fire. Eventually the tandini lands on the roof of the house and, using scratching and whistling noises, speaks to the medium, who interprets the message. Frequently the tandini offers to aid his kinsmen by causing mists and darkness that prevent enemies from seeing them or asks to have his death avenged by killing a particular man. Sometimes the tandini indicates that the fighting needs to be stopped or more people from the kin group will be killed. If the latter, the group sends a message of peace to their enemies to begin compensation procedures. Importantly, though, revenge is often driven by the tandini of the deceased; it causes "kinsmen to 'crave a man' (*akali yiya*) and seek revenge" (Biersack 1991: 257).

As spirit mediums, women have active roles in either prolonging or ending intergroup conflicts. But while conflicts are common in Porgera, the consequences of actually killing someone are profound. Not only does the tandini of a deceased warrior often seek revenge for his killing from his clan members, but the tandini continually seeks revenge on his killer through spiritual attacks. These can occur in a variety of ways but mainly by the killer's failure to observe some prohibition. If a killer sits on the ground on land other than his own group's, the tandini of a man he has killed can enter his body through the ground and attack him. Many killers die from a disease called *akali nono* (eating the man). This occurs when a killer shares food, cigarettes, or drinks with anyone from his victim's social group. The idea behind this is that when the body of a slain man is buried his blood flows out to the boundaries of his group's land. His group members therefore all have a tiny bit of his blood on their hands from working the land. Passing a sweet potato, a cigarette, or a can of Coke to a killer transfers that bit of blood to him and allows the tandini to find him and attack him. In a close-knit society where reciprocity is the norm it is easy for a killer to unwittingly "eat the man" he killed. I unknowingly nearly gave akali nono to a good friend when I served acorns from the pai tree ("bush rice") for dinner one night. I bought them from a woman selling them at a roadside stand. Fortunately we realized that I had bought them from the kinswoman of a man my dinner guest had killed in tribal fighting. (He ate normal rice instead.)

Spirits have a profound effect on Porgeran social and ecological behavior. Large-scale earth renewal rituals that existed until the 1960s, like the rituals at Auwalo Anda and Tuandaka, were tied together into regional ritual networks that linked groups across boundaries of language and culture. In the contemporary era of resource development these same rituals are thought to be responsible for the presence of minerals and petroleum. The spirit world is largely an unseen world; the roads that spirits travel are the numerous underground tunnels and waterways that perforate the porous Porgeran limestone landscape (Biersack 1991: 236). Spirits make themselves manifest at various times, today especially as tandinis of individuals. The fear of spirit attack encourages generosity and sharing through reciprocity at the same time that it fosters secrecy and magic. Warfare and killing are transgressive yet fundamental to the flow of wealth through Porgeran society. Concern over spirit attacks that stem from killing shapes the decisions that households and kin groups make over whom to interact with socially and where to plant horticultural gardens. As mentioned in chapter 4, eating is a powerful metaphor that highlights ideas of reciprocity and the dangers associated with failing to reciprocate.

The chapters on land, people, and spirits demonstrate that linkages between people, the natural and metaphysical worlds, and production are necessary for all life to exist as Porgerans know it. These chapters also stress the indivisibility of the world into Cartesian categories such as "culture" or "nature." Human agency at work in social and environmental spheres is a critical component of the reality of existence. It requires sustained individual and collective efforts to garden, to organize and sponsor marriages and exchanges, and to conduct the rituals that are believed to maintain ecological processes. At the heart of the Porgeran worldview is an ethic of work and the sacrifices that this work requires from the bodies and lands of Porgera. Life for Porgerans is a cycle of work that erodes bodies and lands in the nature of production (and reproduction) but which then reconstitutes them in the rituals oriented toward the production (and reproduction) of nature. People, plants, land, and animals all share the vital life force of ipane that is mediated through spirits. People and animals (but not spirits) also share nembo, or consciousness. In this sense Porgeran ontology is animistic (*sensu* Descola 2013). At the same time, the "sociocosmic hierarchy" of sky people, ancestral spirits, earth mother women spirits, moral topography, and sacrifice characterize Porgeran ontology as "analogic" (Descola 2013). Rather than try to fit Porgeran ontology into a typology, a more critical task is to document how ontologies are transformed by external events and influences. This is not the homeostasis of a 1960s environmental anthropology (Rappaport 1967) but is instead attentive to the fluxes and variations in the human-environmental equation that I explore in more detail in the next part of this book. What is at stake in the formation of a resource frontier? In what ways do the disruptive effects of capitalist resource extraction destabilize local cosmologies, forms of social organization, and ecological knowledges and practice? How do people struggle to maintain meaningful lives and hold to their ontological worlds throughout this process?

Part III

Social-Ecological Perturbations and Human Responses

Part III highlights the complexities of landscape, kinship, and ritual in describing Porgeran livelihoods. Perceptions of place shape subsistence practices and resource management and the appropriate zones in which to pursue these activities; knowledge of kinship and myth provides access to these various zones on different groups' lands, but spirits and past histories of warfare and killing put constraints on mobility. Yet mobility is central to the resilience of this complex system as people need access to diverse land holdings to overcome endemic conflict and climatic variability. Resilience is an important concept for understanding how societies accommodate change because "the adaptive capacity of all levels of society is constrained by the resilience of their institutions and the natural systems on which they depend. The greater their resilience, the greater is their ability to absorb shocks and perturbations and to adapt to change" (Berkes et al. 2003a: 14). One of the key founders of the concept of resilience, C. S. Holling, recently urged researchers to not just focus on highlighting successful examples of resilient systems but to also examine the limitations of resilience, or the "tipping points" at which systems lose resilience.¹ In the following two chapters I explore the challenges to social-ecological resilience. In chapter 6 I focus on ecological disturbances and people's responses to them, and in chapter 7 I examine social disruptions and the ongoing impacts this has on Porgeran society. While I separate these for the purposes of analysis, I reiterate that they are linked in actuality, but at times separation makes it easier to narrow in on deforestation, on the one hand, or income inequalities, on the other.

Holling and Gunderson (2002) offer a metaphor, "the adaptive cycle," to describe the role of disturbance in a social-ecological system (figure III.1a). Disturbances provide important insights over time, as all systems change, but "disturbances have qualitatively different effects on social-ecological systems than do short-term variability and gradual change" (Chapin et al. 2009: 15). An important caveat Holling and Gunderson make is that the adaptive cycle is not a model but a heuristic for thinking about how to understand dynamic change. There are four phases that describe different aspects of the cycle: exploitation, conservation, release, and reorganization. As originally conceived and applied to ecosystems, the exploitation phase is a period of rapid colonization, such as after a disturbance, while the conservation phase is a time of slow accumulation during which energy and material (i.e., natural capital in ecological systems and social capital in cultural systems) are stored. At this point the adaptive cycle is like the pioneer-climax model of Clementsian succession (Clements 1916). However, the shift away from ideas of climax and homeostasis in ecological understanding requires two more phases. The first of these is release or "creative destruction" (Holling and Gunderson 2002: 34), in which energy and materials are released through agents such as forest fire, drought, pests, conflict, or shifting cultivation. The second is reorganization, in which material is reorganized to become available for the next phase of exploitation. The reorganization phase is the point at which novel elements can be introduced to allow for changes in the system, biotic legacies allow the system to reorganize in a way similar to the previous system, or leakages of matter transform the system to one that is often less productive and organized than the previous system (Holling 1973). Importantly, systems do not necessarily move through the sequence of phases in a similar way, and the adaptive cycle can also be used to understand change in social systems (figure III.1b and c).

A critical property of the adaptive cycle is resilience, which fluctuates through the different phases. The reorganization and exploitation phases exhibit high resilience, while the conservation and release phases are periods of low resilience. The movement from the exploitation to the conservation phase is a "slow, incremental phase of growth and accumulation" that is fairly predictable, while the shift from the release to the reorganization phase is rapid and "highly unpredictable and uncertain" as to the outcomes (Holling and Gunderson 2002: 47). The

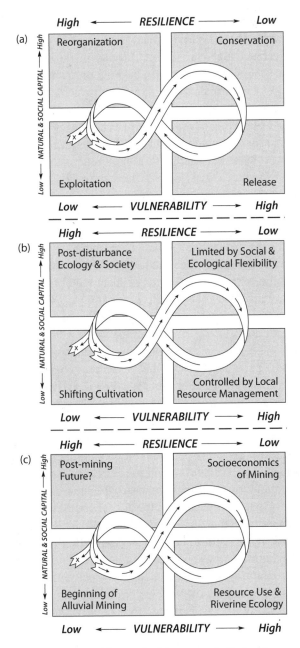

Figure III.1. The adaptive cycle (a). Length of the arrows indicates the speed of change from one phase to another. X represents the point at which elements are lost or incorporated into the cycle. The time before mining in Porgera (b) and the contemporary situation are explained in the text (c). Adapted from Gunderson and Holling 2002.

conservation phase is where most systems typically spend their time and is also when there is the greatest likelihood for vulnerabilities to disrupt the system (Chapin et al. 2009). Studies in a resilience framework suggest that management strategies based on traditional ecological knowledge "put the brakes" (Berkes and Folke 2002: 125) on transformative changes in the release phase of the cycle by creating small-scale disturbances to which people and their environment can readily adapt. As well, *social memory*, which I define as the suite of memories that are generated from local knowledge and experiences that shape current practices and ideologies that people share among one another, serves as a repository for mitigating disturbances (see chapter 6).

I argue that both the ecology and society of Porgera can be viewed heuristically by using the adaptive cycle in a temporal framework. In the case of the pre-mining social-ecological system in Porgera (as seen in figure III.1b), vulnerabilities were reduced in the conservation phase through mobility and flexibility such as discussed in chapters 4 and 5. Nature spirits controlled the release phase by limiting extensive resource harvesting as a consequence of reprisal. Moreover resource management practices such as maintaining multiple garden sites helped to put the brakes on release in the subsistence cycle (see chapter 6). In essence traditional ecological knowledge and practice in Porgera worked to keep the social-ecological system between the extremes of low and high resilience and high and low vulnerability. With the onset of mining, as I describe in chapters 6 and 7, resilience has been decreased while vulnerabilities have been increased through changing socioeconomic patterns and pressures on traditional resource management. Mining development has introduced vulnerabilities into social-ecological systems in Porgera by reducing the flexibility with which social groups form and the mobility of people to access different lands. Mining development itself is a complex social-ecological system that exhibits different phases of the adaptive cycle simultaneously (see figure III.1c). Initially the early years of alluvial mining were highly resilient (the exploitation phase) in terms of allowing a diversity of people to access the goldfields. The riverine system is clearly undergoing profound ecological change in the release phase, with 17,000 metric tons of waste being dumped into the river daily, heralding as yet unknown consequences for the future (Shearman 2001). Socially, mining benefit payments are in the conservation phase, with money being accrued and hoarded by small groups of Special Mining Lease landowners. As I discuss in chapter 7, this is a key source around which social tensions erupt and increase the vulnerabilities associated with mining development.

There are numerous examples both socially and ecologically in which mining has impacted the resilience of livelihoods in Porgera, which articulate in some ways with James Scott's (1998) concept of simplification. Whereas Scott saw an authoritarian, coercive state and a prostrate civil society as necessary conditions for explaining the failure of many large-scale development schemes around the world, PNG exhibits neither of these political realities. The Porgera case does support simplification as dependent on the administrative ordering of nature and society and a "high-modernist ideology" (Scott 1998) based on concepts of scientific progress, rationality, and increasing production. While Scott defined simplification as a form of statecraft in which societies are made legible through censuses, cadastral surveys, and other forms of standardization (see also Li 2007), the sense in which I use the term is less overt, as something emergent in most capitalist, rationalistic systems. As an example of this, Rappaport (1971: 130), characterizing the difference between the multicropping of New Guinea horticulturalists and the monocropping of Western agricultural systems, argued, "The anthropocentric trend is in the direction of simplicity rather than complexity, of fragility rather than stability." Replacing fragility with vulnerability and stability with resilience captures the sense in which I deploy simplification. Put bluntly, simplification erodes resilience.

As noted, processes of simplification lend themselves to promoting conflicts over access to resources. In a sense violence is a response to neoliberal policies and simplification. This feature of their behavior makes linking resilience frameworks with political ecology an important undertaking. In providing a framework for political ecology analysis, Raymond Bryant (1992) highlights three critical areas of investigation: the contextual sources of environmental change, conflicts over access to environmental resources, and the political ramifications of environmental change. I address these in the following chapters but am also attendant to the social and cultural ramifications of environmental change (see Moore 1993). Additionally, building on concerns over the lack of ecology in political ecology (Vayda and Walters 1999; Walker 2005), I explore the ecological implications of environmental transformation to comprehensively link dialectical social and environmental processes of change.

A final theme that I address in chapters 6 and 7 is the idea that the land is ending. In chapter 6 I discuss ecological changes and declines that Porgerans perceive to be impacting their environment. Importantly, ecological declines are linked to moral declines in society that are the topic of chapter 7. These linkages between ecological and societal decline serve as a perfect example of the inseparability of the social and the ecological in social-ecological systems.

*ecological disturbances and
people's reactions to them

CHAPTER 6

Ecological Perturbations
and Human Responses

Disturbance and Social Memory

In this chapter I examine several ecological disturbances—volcanic eruptions, droughts caused by El Niño, climate change, and demographic
changes and pressures on forest resources—to examine resilience and
vulnerability in human responses to these perturbations. Long-term social memory is an important component of resilience because of its ability
to serve as a buffer against change (Berkes et al. 2003a). In her study of
the oral history and social memory of glaciers in the St. Elias Mountains of Alaska and the Yukon, Julie Cruikshank (2005: 11–12) writes that
"changing landscape features, such as fluctuating glaciers, have provided
imaginative grist for comprehending and interpreting *shifting* social circumstances." In coupled social-ecological systems such as Porgera and
elsewhere, ecological changes and social transformations are perceived as
inseparable. Social memory, local knowledge, and regional environments
are dynamic and transform in order to accommodate novel processes.
However, there are limits to resilience when people migrate into the
high-altitude rain forest (aiyandaka) to be nearer mining development.

In my first exploration of social memory, ecological perturbation,
and human response, I turn to a mid-1600s volcanic eruption on Long
Island, off the northern coast of PNG, that had the same explosive
force as the 1883 eruption of Krakatau (Blong 1982: 4). The resulting
ash cloud spread west, covering much of the highlands as far as the

Q 1

Strickland Gorge (Blong 1982: 155, Fig. 56). As Russell Blong documents in his book *The Time of Darkness*, this event has endured in the oral histories of highland Papua New Guineans for over three hundred years in a series of legends about a *taim bilong tudak* ("time of darkness" in Tok Pisin) that span multiple cultural and linguistic groups. Among the Ipili and Huli, who lived at the most extreme western edge of the ash fall, the time of darkness gained a cosmological significance not present in other legends, even though the ash layer would have been but a few centimeters deep (Ballard 1995: 128; Ballard 1998: 68–69; Blong 1982: 51, Fig. 24). In both of these groups the time of darkness was linked to ecological destruction followed by renewal and extraordinary abundance as the fertile ash revitalized soils. Cultural proscriptions were put in place to prepare for the next time of darkness and orally passed down from one generation to the next. If these proscriptions were not followed, ecological and cultural apocalypse would ensue (Glasse 1995).

Ipili speakers call the time of darkness *ungi* (a contraction of *undupi*, "dark," and *ngi*, "in the time of"). Thunder, lightning, and the gathering of dark clouds on the horizon mark the onset of ungi (Meggitt 1973: 28). In 1999 Muyu Yakati described the coming of ungi and the attendant cultural proscriptions:

> When we heard that ungi was about to come, we were told by our ancestors to build a house and go inside it for protection. The house we built didn't have pandanus leaf roofing. We were told to use cane grass [*Miscanthus floridulus*] to roof it. Ungi is like cement [in its powdered form]; it falls like dust from the sky and then hardens on the ground. When something falls on it, though, it will break. Our ancestors told us that when ungi comes we must cut cane grass and put it wherever our gardens are flat. Gardens on slopes would be all right as the ash would eventually wash off in the rain. Afterward we could lift the cane grass coverings off of our gardens and the ungi would not destroy our crops.
>
> When ungi is coming we put firewood and food inside the special house we have built; we build a toilet inside because we know we can't go outside during ungi. A child that has no siblings, however, can go outside during ungi because ungi knows that the child is just one, and ungi won't kill the child. Any boy that had no siblings was designated to be the child that could go outside during ungi and get more firewood or water if it was needed. Lendeapuni Tole is the last man that was given this designation; he's an old man today. Families

that had more than one child had to stay inside for five days, then the first born child could go out; the next day the second born child could go out; the third day the third child could go out; and so on.

When ungi comes, we were told to abandon the sexual segregation rules. Men and women slept in one house, all mixed together. Everyone stayed inside for five days. All night and day we were told we would hear the ungi falling and lots of noise like thunder. On the fifth day, the ungi would end. On the fifth day, everyone comes out, and it looks like ashes from a fire have covered everything. We go to the flat places and remove the cane grass coverings that protected our gardens. Where the ash has fallen everything dies, but afterward vegetables will grow huge and the land will be more fertile. Pandanus trees will produce tons of nuts. There will be large lumps of this ash, and we were told to get branches from trees and roll them to the sides of the gardens.

At Umambo, Kenola's grandfather, Yongolo's father, was born the last time ungi came; his parents named him Undupi. [As Kenola was in his sixties, this would have been around the late nineteenth century; however, according to other informants, Undupi was Yongolo's grandfather, perhaps pushing back the event to the mid-nineteenth century.] The next time that ungi comes the ground will end, we were told. Helape and Kelape will come up [to Porgera] and we will all die. So you would think that when ungi comes it will be good because the ground will become fertile, but Helape and Kelape will let go of the yu kimbuni [earth joint] and so will the man at Tuandaka. They will all meet at Aiyapo Swamp [just south of Porgera behind Mt. Asienda], beat their drums and dance, and then the ground will end. (field notes, January 1999)

Tisapa Yawe, another elder, recalled:

Before my grandfather's time there was a man named Undupi. Undupi's father's name was Pu. Pu's wife was pregnant with Undupi, and a stone [ana] fell from the sky. It was like ash. After this happened it got dark—the sky darkened. During this time, no one went outside, everyone stayed inside their houses. While they were sleeping, pythons came inside the house. It was good for the people to kill them and then eat them. It would have been bad if they had killed them and thrown them away. Everyone stayed inside for four days. On the fifth day the light returned and we could go outside. (field notes, January 1999)

Wambe Yawin added to the list of proscriptions: "The old men used to tell us that we should always keep some pandanus stored away for when ungi comes. Ungi was a white dust. They told us that we had to get cane grass and cover up our gardens that were on flat ground, gardens on slopes would be all right." Kamino Kipako, a western Enga man from Wailya, said, "Our fathers told us to always garden on slopes. If we made gardens on flat places, ungi would destroy them."

Ethnographers among the Huli (Ballard 1995, 1998, 2000; Frankel 1986; Glasse 1995; Goldman 1983), explain that the "earth magic" (*dindi gamu*) rituals (detailed in chapter 5) were intended to bring about the time of plenty that would follow the time of darkness. Today, among the Ipili, as noted in Muyu's narrative, the next occurrence of ungi will not be followed by a time of plenty but will foretell the end of the world. While it will eventuate in spirits "letting go" of the earth joints, culminating in world destruction, it has also been syncretized with Christian concepts from the Book of Revelation. Muyu said, "My father didn't tell me all the white men and their stuff would come someday. When you white men came it scared us because our fathers didn't tell us about it. The last time ungi came, big stones fell from the sky at Mt. Tongolo and Liwiya. When ungi comes our ancestors told us we'd die. Now the Bible, planes, axes, and white men have come. Your Bible tells us that the ground will end. Our ancestors told us ungi would make the ground end too." Tisapa too linked the ungi occurrences to Christian belief and practice: "At the place where you white men live, you killed Jesus, and God was mad and sent the ungi down and made the ground go dark. This was seven generations ago. Pu fathered Undupi, Undupi fathered Kuela, Kuela fathered Yongolo, Yongolo fathered Kenola, Kenola fathered Balus, Balus has a child—that's seven. Now we Seventh Day Adventists usually worship on every seventh day. Our ancestors told us that this ungi would come again. So this time of darkness will come again too. It will come after seven generations [i.e., any time now]."

The Ipili and Huli, who lived farthest from the effects of the volcanic ash fall, developed one of the most significant cultural responses to it (see Blong 1982: 130–31). As Robert Glasse (1995: 58) argued for the Huli, the time of darkness articulated with Huli concerns over an "unstable environment" and agricultural practices that produced "fluctuating yields," resulting in "a boom-or-bust economy." Ideas and practices associated with the time of darkness reflect cultural concerns about the need for human intervention into ecological processes and the activities of spirits that hold the land together. Cultural responses to an event that happened over three hundred years ago demonstrate the longevity of social memory for environmental perturbation in groups that live in shifting and uncertain environments.[1]

Responses to Climatic Disturbances in the PNG Highlands

Following a severe drought in PNG in 1972, Eric Waddell (1975) published an article in the journal *Human Ecology* that highlighted a suite of differing responses that Enga speakers made in response to climatic disturbances during that event. Severe droughts in the PNG highlands are closely linked to very strong El Niño–Southern Oscillation (ENSO) events, when the rise in surface atmospheric pressures over the western Pacific (the Southern Oscillation component of ENSO) prevents nighttime cloud cover and creates the necessary conditions for frosts on lands higher than about 2,200 meters above sea level (Brown and Powell 1974).[2] The staple sweet potato is detrimentally impacted by frosts that freeze the aboveground vines, destroying them and preventing further tuber development in the ground. Between June and October 1972, Waddell (1975: 250) documented over thirty nights of frost above 2,300 meters and individual frost events as low as 1,650 meters, "all this in a region where the subsistence food complex is almost exclusively of tropical lowland origin and therefore not frost tolerant."

The Australian colonial government's response to the drought and frosts was a massive food replacement program, of which Waddell was critical because it assumed that PNG highlanders had no mechanisms to cope with climatic disturbances on their own. Contrary to the view that Enga were victims of a sometimes harsh and variable climate, Waddell (1975) argued that they had a series of structured responses whose magnitude varied with the intensity of the climatic disturbance. At the level of everyday practices, Enga plant their crops in raised mounds about 0.5 meter high that increase the temperature by 2°C at the top of the mounds compared to ground-level measurements (252). Mounds are made from old sweet potato vines, weeds, and other nearby vegetation, and the decomposition of this mulch in the mounds raises soil temperature by about 1.2°C. Porgerans cultivate their crops the same way even though most of Porgera's agricultural fields lie below 2,300 meters. They give three reasons for doing so: (1) albeit very infrequently, frosts can strike even the lowest altitude gardens at around 1,600 meters, and mounding helps prevent crop loss; (2) with the high clay content in the soil and the heavy rains, mounding helps to keep the soil from becoming waterlogged; and (3) the decomposition process that occurs in the mound, especially when pandanus leaves are added, provides increased nutrients to the crops planted in the mound. In Porgera mounding thus has other benefits besides protecting crops from frosts (see also Waddell 1972: chapter 5).

For the Engans dwelling at the highest altitude, where frosts are annual occurrences, mounding itself is ineffectual and other ecological and social mechanisms are important for ensuring resource availability. Waddell (1975: 255–56) referred to these mechanisms as "three distinct levels of coping with the problem": the "local, intraregional, and extraregional." The local-level response is to plant gardens in two distinct ecological habitats: valley bottoms and the lower slopes of adjacent ranges. While the former contain more fertile and productive soils, they are also more vulnerable to frost. Planting a few gardens on the less productive slopes provides some livelihood insurance during minor frost events. The intraregional level of response is a variation on a theme of the local-level response. Gardens are maintained up to a day's walk from the main homestead in different areas to ensure that if localized frosts destroy a few of a household's gardens, others will likely have escaped damage and will still provide food. Occasionally massive frost events (in the early 1920s, 1941, and 1972; after the publication of Waddell's 1975 article also 1982 and 1997–98) destroy all of the high-altitude gardens, and the extraregional level of response is enacted, which takes the form of a massive out-migration to lower altitude areas. Porgera is one of the destinations of out-migration during these events (see Map 3.1).

As Waddell (1975: 262) emphasizes, out-migration is not a haphazard strategy but the result of a long-term series of reciprocal exchanges between potential hosts at lower elevations and potential guests at higher elevations. In Porgera the Wage and Kandep areas (the high-altitude areas in Waddell 1975) are renowned for the abundance, size, and taste of pigs raised there. During the past century of severe frosts Porgerans have willingly welcomed high-altitude Engan dwellers to live among them in order to access this porcine wealth. Since living memory, the hosts and guests have intermarried, further strengthening the connections between high-altitude Engans and Porgerans beyond just those of exchange partners. In fact many of my "Porgeran" informants are the offspring of past Engan climate migrants and Ipili hosts. When I first arrived in Porgera in late 1998 there were still a number of families from the Wage and Kandep areas living on Porgeran hosts' lands due to the frosts stemming from the 1997–98 El Niño. Over the next fourteen months I attended several marriage ceremonies between Engan migrants and Porgeran hosts, which reproduced the very social connections that accounted for a large number of my Porgeran friends and informants.

Kainu Muyaka (figure 6.1) was a young girl when her father migrated from the Kandep area to Porgera following the 1941 drought and frost.

Figure 6.1. Kainu Muyaka (*right*) with her daughter Julie.

pu-ipu - social connections resulting from frost event good ex.

During this time Kainu's elder sister met and married Timbapu, the man who was later involved in the killings at Tipinini in the 1940s cults established after the frost (see chapter 2).[3] After a time Kainu's father took her, her mother, and her other siblings back to Kandep, but within a short while Kainu returned to Porgera because she missed her sister. With two daughters in Porgera, Kainu's family started to pu ipu (go and come): "We came and went between Kandep and Porgera all the time." In 1999, when I was interviewing Kainu, her husband and eldest son had been killed in tribal fighting, and she was living with her sister (father's brother's daughter), Lombeama. Lombeama's migration story, similar to Kainu's, began with Lombeama following Kainu to Porgera and eventually marrying a Porgeran man. "We always follow our sisters," concluded Lombeama. In 1972 Paul Wohlt (1989: 229) also documented frost-caused migrations from the Wage area to Porgera and other lower areas and found that approximately "half of the families went to the same place to which the husband or his father had gone in 1941." Wohlt described the scene in the host valleys as "a time of increased socialization," with courting, dances, and "continual coming and going" between the high-altitude homelands and the host areas.

The areas to which high-altitude Engans migrate are not themselves immune from the severe climatic perturbations that initiate migration. In 1997 many Porgerans lost crops from ENSO-caused frosts, albeit not at the same levels of loss as high-altitude dwellers. The ensuing drought, though, caused even more crop losses and forced people to travel long distances to gather water as the numerous small springs dried up and only the major rivers maintained a slight trickle of water. In 1955 frosts decimated the food supplies in the upper Porgera valley, and 50 percent of the population migrated to Tipinini and areas outside of Porgera for food (Laiagam PR #3 1955/56).

Thus social memory in both the long term (several generations) and medium term (one or two generations) articulated with ecological perturbations and cultural responses. In the next section I turn to the short term and quantitatively analyze the relationship between ENSO patterns and climate in Porgera and local perceptions of a changing climate. A critical concern is exploring how well social memory matches climatic data.

El Niño–Southern Oscillation and Climatic Variations in Porgera

When I was in Porgera between 1998 and 2000, I didn't collect or have access to climatic data, but what I did gather were numerous oral stories about climatic changes and severe weather events (see also Jacka 2009).

Upon analyzing my climate oral histories, I became interested in how to reconcile local perceptions of climate change with data on rainfall and temperature. I also wondered how environmental anthropologists could use climate data to conceptualize severe disturbances and explore human responses to these disruptions. Oral history of severe climatic disruptions points to the years 1941, 1972, 1982, and 1997 as periods in which extreme droughts and frosts necessitated the "extraregional" responses discussed earlier,[4] in which Porgerans hosted climate refugees from the high-altitude Enga areas to their southeast. As mentioned in chapter 1, Porgera is characterized by aseasonal, albeit highly variable rainfall, with the lowest monthly rainfall registering just over 50 millimeters and the highest just under 600 millimeters. As also mentioned, Porgeran perceptions of seasonality are expressed in the terms kingi (wet) and panyu (dry), which are not linked to any particular time of year but are instead used to depict the dominant weather pattern currently. In order to compare the relationship between ENSO events and rainfall in Porgera, I overlaid the ENSO index as calculated by the National Oceanic and Atmospheric Administration's Climate Prediction Center onto a rainfall index I created with monthly precipitation data from 1974 to 2011 I received from PJV (figure 6.2).[5]

In figure 6.2 positive standard deviations (SD) are periods when more rain fell than normal, while negative SDs indicate that less rain fell. As can be seen, the strong El Niños of 1982 and 1997 (>2 SD) are associated with severely reduced rainfall in Porgera. In general very strong El Niños mean less rainfall; however, there are periods (1990, 1993, and 2010) when this relationship is not as strong.

While ENSOs and droughts in the western Pacific typically generate the most attention from the international community and subsistence horticulturalists, more accurately droughts and extreme wet periods in particular sequences result in the lowest yields of sweet potatoes. As Robert Bourke (1989) has shown, the most severe crop losses occur after periods of extreme rainfall that are followed by droughts, even minor ones. Wet periods that occur in the three to ten weeks after initial planting reduce tuberous root development and limit development to the top 15 to 25 centimeters of the soil profile of a mound. A drought after this period of time, when the tubers are developing, greatly reduces output. And of course a drought accompanied by a freeze completely kills the plants at this critical stage of tuber development. Bourke argues that droughts alone fail to cause famines and that extreme waterlogging of the soil is a better predictor of crop failures. In general, oral testimony by Porgerans highlights that extreme climatic events of too much or too

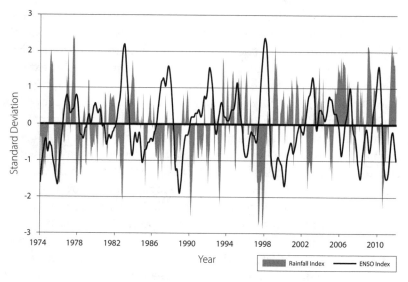

Figure 6.2. Rainfall index overlaid with El Niño–Southern Oscillation index, showing the relationship between strong El Niños and droughts in Porgera.

little rain impact their livelihoods, but nevertheless nearly all of their oral history focuses on droughts.

In an attempt to quantify both the duration and the intensity of wet and dry periods, I modified Allen's (1989) dry season index for PNG and derived three climate types (normal, wet, and dry) from the rainfall index to determine what constitutes an extreme climate event (table 6.1).[6] The wet periods are indicated by the gray bars and the dry by the white bars; both show the duration of the climate anomaly, while the number indicates the summed standard deviation index score and hence the intensity of the climatic anomaly (the ellipsis indicates continuation from the previous year). As seen in table 6.1, between the beginning of 1974 and the end of 2011 there were seven extremely wet periods and eight extremely dry periods. Of the eight dry periods, only three, 1982, 1997, and 2002, overlap (although not perfectly) with official El Niño periods according to NOAA's Climate Prediction Center.

In interviews conducted in 1999, people often mentioned the frosts and droughts of the 1982 and 1997 events and how they lost crops and hosted Engan migrants. Interestingly no one in Porgera ever mentioned that there was any drought during the 1990 and 1993 periods—despite the intensity of the 1990 drought surpassing the 1982 drought. Evidence suggests that proceeds from the mining development and the Mt. Kare gold rush provided enough of a buffer for subsistence during these two

Table 6.1 Extreme wet and dry periods in Porgera

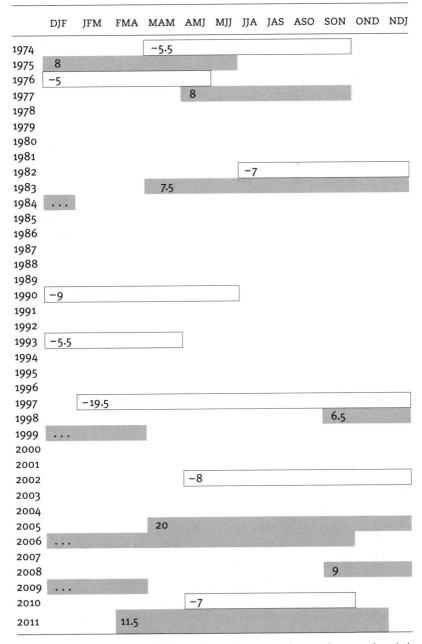

	DJF	JFM	FMA	MAM	AMJ	MJJ	JJA	JAS	ASO	SON	OND	NDJ
1974			−5.5									
1975	8											
1976	−5											
1977				8								
1978												
1979												
1980												
1981												
1982						−7						
1983				7.5								
1984	...											
1985												
1986												
1987												
1988												
1989												
1990	−9											
1991												
1992												
1993	−5.5											
1994												
1995												
1996												
1997	−19.5											
1998										6.5		
1999	...											
2000												
2001												
2002					−8							
2003												
2004												
2005				20								
2006	...											
2007												
2008										9		
2009	...											
2010					−7							
2011			11.5									

The letters DJF, JFM, and so on, indicate the weather averaged over a three-month period. For example, DJF is the period December, January, and February.

periods as people had the ability to purchase foodstuffs. The most significant dry period, in 1997, preceded the official onset of the 1997–98
El Niño, which began in May, but similar to the drought during the 1982
El Niño, rainfall returned to normal in Porgera several months before
the official end of El Niño in the western Pacific.

Porgerans told me that a few months into the wetter season (around
January) is a time of famine in the valley that is alleviated only by the fact
that nut pandanus season begins about this time. Famine in this context must be understood from its cultural perspective in that there are
often many kinds of food available, but not the beloved staple of sweet
potato (cf. Sillitoe 1996: 79–86). Occasionally, however, there are times
when an overabundance of rainfall impacts Porgeran subsistence. The
wet period data correlate closely with Porgeran perceptions; five of the
seven wettest periods occur around the beginning of the year. I should
note that wet periods in Porgera are not linked to La Niña events in any
significant way; there is some overlap, but the data are not as consistent
as strong positive ENSO values and severe El Niños that remain drivers
for dry periods in Porgera.

Temperature data from 1988 through 2003 from Porgera that I received
from PJV show less variability than data for precipitation. As would be
expected in the aseasonal tropics, there is little month-to-month variation and very small monthly standard deviations. Average monthly
highs range between 19.7° and 20.6°C, while average monthly lows range
from 11.3° to 12.3°C. The general pattern of variability between lower
and higher temperatures is that the wet season in Porgera typically has
higher average temperatures than the drier season. For example, the
lowest mean monthly temperature (9.8°C) occurred in the middle of the
1997 El Niño in September.

While there is of course local recognition of climatic variability, there
are also very strongly stated opinions that the climate is dramatically
changing in Porgera due to the impacts of mining development and
Christianity. During my 1998 to 2000 research period, people stressed
that the climate in Porgera was becoming hotter and drier. Birds that
lived only in the lower rain forest could now be seen in the andakama.
Crops, such as bananas, that had never been able to grow in the upper
rain forest now could. To examine whether this was represented in the
climatic data, I calculated cumulative differences from the mean for
both rainfall and temperature. In figure 6.3 downward slopes indicate
negative deviations from the mean of the three-month running average rainfall, while upward slopes indicate positive deviations from the
mean. The data in this example show that between November 1977 and

Figure 6.3. Cumulative deviations from mean rainfall, 1974–2011.

October 1998, there was a twenty-one-year general decline in mean rainfall, with the exception of a three-year period between May 1993 and May 1996. Since the end of the 1998 El Niño the climate has been wetter than average over the long term. This demonstrates the importance of paying close attention to local knowledge about climatic patterns. In early 2000, at the end of the time I was collecting data on perceptions of climate change, there had been a general decrease in rainfall over a two-decade period. Statements that the climate was "becoming drier" correlate closely with the rainfall data. However, in 2006 people were complaining about waterlogged soils and decreased crop production in their gardens. Given that they had just come out of the longest, most intense wet period (nineteen consecutive months when rainfall surpassed 0.5 standard deviations each month), this is not surprising.

Figure 6.4 is a more complex picture of temperature deviations. From January 1988 to September 1998 the maximum temperature shows a warming trend; this is especially prevalent from August 1994 to September 1998. The minimum temperatures, however, show a reverse trend, with the minimums generally decreasing from January 1988 to November 1997. In essence, local statements are partially correct. It was indeed getting hotter in Porgera from the perspective of the late 1990s, if one was talking about the daytime temperatures. Nighttime temperatures were becoming colder, though. Since people spend little time outside in the early dawn hours due to fear of nighttime spirits, I can only assume that their perceptions of a warming trend were based

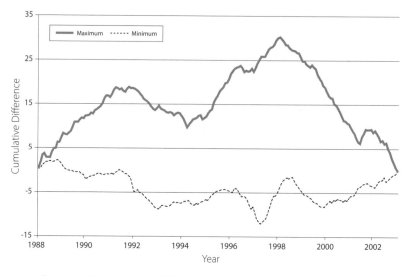

Figure 6.4. Cumulative deviations from mean temperature, 1988–2003.

on afternoon temperatures. Additionally climate interviews coincided with a surge in millenarian and apocalyptic Christian narratives of the world ending in the year 2000, and many interviewees mentioned "the fire in the sky" that God was bringing to usher in the Second Coming and the impending time of tribulations (see also Stewart and Strathern 1997).

The 1997 El Niño and drought caused considerable social and ecological disruption. Crop losses in Porgera were severe, and with the in-migration from high-altitude Enga areas, food aid (mostly rice) was provided by the mine and government. With the restriction of mining wealth and the demise of the Mt. Kare gold rush, there was also less money available for purchasing food (unlike in the 1990 and 1993 El Niños). In the daily household context, women especially remember this as a troubled time not just due to the loss of garden crops but in terms of the difficulty of obtaining fresh water. What is normally a one- or two-minute walk to one of many nearby springs turned into an hour or longer trek to a river, with a return trip carrying heavy plastic water jugs. Atmospheric haze from the forest fires in Indonesia caused the rising and setting sun and moon to turn blood red, which articulated with growing concerns over the Christian apocalypse, widely believed to coincide with the year 2000. An earthquake and eclipse in early 1998 only intensified these fears of apocalyptic transformation.

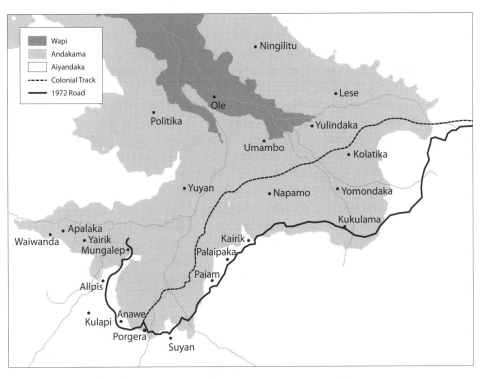

Map 6.1. The southern portion of the Porgera valley.

Population Shifts and the Political Ecology of Migration

Migration, both internal as Porgerans have moved to be near the highway and external as outsiders have immigrated into Porgera, has severely impacted the availability of forest resources, especially trees, for firewood and building materials. The relationship between forest degradation and population change is not linear and cannot be explained by simplistic neo-Malthusian assumptions of carrying capacity and "balance of nature" arguments (Scoones 1999). Rather degradation is constituted socially through land-tenure decisions and land-use practices related to local perceptions of resource scarcity, and it is constituted ecologically by different forest-regeneration processes at various altitudes of extraction. Thus the issue of forest degradation and human population growth is best addressed by political economy and ecology equally.

In the colonial era the Australian government constructed first a footpath and then a motorcycle track into Porgera through the most populous parts of the valley in the midaltitude andakama ecozone (Map 6.1). Then, in the early 1970s, the colonial government built a vehicle road

Ecological Perturbations and Human Responses | 177

in the high-altitude aiyandaka, which bypassed the main settlements and was several hundred meters higher in elevation than where most people lived. The decision to build along this route was a process in simplification: while no communities lived near the road, the route was much flatter than the previous track. When the road was completed in 1972, initially few people moved to be near it as it lay entirely within the high-altitude forest that was the abode of the sky people. With increasing mining exploration, however, people started moving closer to the highway. Matokas Elata, who moved from lower down in 1981 or 1982, remarked that "people didn't use to live up here, it was just a place for hunting marsupials." When mining development became a reality in the late 1980s, though, people realized that moving up to the highway had various economic benefits; any land destroyed during the widening of the highway to accompany trucks bringing supplies to the mine would be compensated by PJV.

In 2006 Jerry Loa recounted to me the migration of the Bipe from their original hamlets in Umambo to Kairik:

> In 1988 we were all still living down at Umambo, and around this time we heard rumors that PJV was going to build an airstrip at Kairik. All the big men told everyone to move up to Kairik. Our fathers and grandfathers used to come up to this area to gather wild pandanus and hunt birds and marsupials. The big men told us to go claim the land where our fathers and grandfathers had built bush houses for hunting.
>
> So we all moved up to Kairik and started chopping down all the big trees, building houses, and clearing land for gardens. This was hard work; before, this was just big bush. So we cleared this area, and then PJV bought a piece of land from us for the airstrip. The people who moved up first got more money from PJV and were able to build metal houses [haus kapa] with their money. Now this makes eighteen years that we've lived here.

There are stipulations on how much developers have to pay for the loss of land to landowners. Land with improvements, such as pandanus trees, gardens, and houses, receives substantially more than unimproved land. The people who moved up first and made the most improvements received the greatest share of benefits. Jerry wistfully remarked that his father had not been one of those to make the initial migration. I asked him if he ever regretted moving up to the highway, as firewood was scarce and gardens failed to produce as well lower down in the valley. Jerry replied, "It's not as if my children are going to spend their lives up here. There's not enough land, and everyone lives too close to each other. When the mine finishes, we'll all go back down and live at Umambo, where we have plenty of land."

The two months I spent living with the Bipe at Apone Camp in Kairik provided considerable insight into the difficulties of living at 2,300 meters above sea level. Other places I had lived that were around 1,800 meters, such as Kolatika and Wailya, had abundant food and firewood resources. In contrast, along the highway wood sellers were offering bundles of four six-foot-long pieces of split firewood for K4 (US$1.33). One of these bundles was sufficient for about two days' worth of cooking, meaning that households without access to firewood had to spend on average about K60 per month, which in many households would be as much as their entire monthly income.[7] What intrigued me about the firewood sellers was that in 1999 there hadn't been any in the eastern Porgera valley; the lack of accessible firewood was a recent phenomenon in 2006. Around the mine, however, firewood had been scarce as early as 1993 (Banks 1999), when a survey of ninety-six households found that 29.2 percent had "short" supplies of firewood and another 40.0 percent had to buy it. People's declarations of a firewood crisis in the eastern Porgera valley seemed somewhat paradoxical, however, as looking down the valley one could still see vast unbroken tracts of rain forest.

The situation with food was somewhat clearer. Even in 1999 gardens in the highway communities were well known to be less productive than andakama gardens lower in elevation. The combination of increased precipitation in the higher elevations, waterlogged soils, and colder temperatures prohibited ample tuber development in the sweet potato mounds. With greater population density along the highway, people couldn't compensate by increasing their garden area because most of the arable land was already in use. As a response to lower yields in the higher elevation gardens, nearly every household maintains gardens at lower altitudes in areas where they formerly lived. Households that are unable or unwilling to keep gardens at lower altitudes must rely on purchased food, especially rice, to supplement their diet. Since most households had access to garden lands lower in altitude, why did they not also have access to trees for firewood?

Tree management practices in Porgera are a part of land-tenure norms. As mentioned in chapter 4, planting trees on a piece of land is the way a person makes a claim to "owning" the skin of the land. Once a garden is cut from forest, the owner of the garden will plant casuarina (*Casuarina oligodon*) and pandanus trees in the plot to mark his or her claim. Casuarina trees are one of the most common and economically important trees. They are the most important firewood species in the highlands (Bourke and Harwood 2009: 392) due to their long burn time, high heat output, and low ash production. Casuarina also fix nitrogen in the soil, which is why people plant them in their gardens. Casuarina and pandanus trees also serve as mnemonic devices to remind the original owner's children and other people in the area of the location of a person's past gardening sites—as children should garden where their parents previously have.

Prior to mining, land was considered abundant and people few. There were few if any restrictions on landownership and people claiming land as their own. Recruitment of new group members was an ongoing concern, and outsiders were always welcome to garden and live on a clan's lands. In the mining era this has started to change. Outsiders (*epo atene*) are still sought after as people almost always worry that their group size is too small. Garden lands are provided for outsiders, but letting these outsiders then claim these lands by planting trees on them is another matter. In the eastern Porgera valley, outsiders were initially allowed to cut forest trees in primary and secondary forests for firewood. By the early to mid-2000s, however, clans started to worry about forest degradation and to create forest reserves called *ima* that only *tene* clan members (descendants of males) could access. In essence two intertwined forces are causing perceptions of scarcity of trees for firewood: outsiders are restricted from entering forest reserves, and outsiders are unable to plant casuarina trees (which is what people normally do) due to concerns over their trying to make claims on land that is starting to be perceived as scarce. The scarcity of firewood and building materials has also led to conflicts as people have cut down trees that don't belong to them.

Another factor contributing to firewood scarcity is the location of where people are migrating to, which is primarily in the upper rain forest zone along the highway. Prior to the beginning of alluvial mining in the 1950s and 1960s, reports suggest that most people lived in the midaltitude *andakama*, but there are conflicting data on how populated the upper Porgera valley was. In March 1939, a few days after discovering Porgera's gold, John Black writes of the upper Porgera, "Tons of food, gardens and population," and notes that he was "surprised at large population hereabouts" (Black MS 8346/3/9). However, a few months later he describes a "smallish population at Yairik, at least a little less than at" Mungalep (Black MS 8346/3/8). A decade later, in early 1948, the patrol officer reported, "Few natives live near the River Kogai and they manage only a bare existence" (Wabag PR #2 1947/48: 3). Another patrol officer supported this: "The Ibri [Ipili] people in the Kogai River area were friendly but had little food for sale . . . [as they] are gardening on a blue shale which is very unproductive" (Wabag PR #3 1947/48). In late 1948 a different patrol officer reported that, with the introduction of steel tools, large tracts of forest were being cleared and new gardens being prepared (Wabag PR #3 1948/49). In December 1948 a government geologist wrote that there were "no natives or gardens" along the Kogai River but that the hamlet of Apalaka had "open hillsides and gardens," indicating settlement nearby (Wabag PR #4 1948/49). Hiking out of the Porgera valley along the route of the present-day highway, the accompanying patrol officer noted that the route was through "dense bush . . . with no breaks, natives, or gardens seen" the entire day. Again in 1949 the patrol officer reported "an appreciable increase in the number of people" from Ole to Mungalep. Standing at Mungalep and looking north, "one can see with the naked eye twenty five settlements of two or more houses on the northern slopes of the Kaiya River and thirteen on the eastern slopes of the Porgera River" (Wabag PR #1 1949/50). In summary, perceptions of population in the upper Porgera were as varied as the people who came through the region.

The first census was conducted in Porgera in October 1957 and counted 2,020 people, 294 in the upper Porgera at Mungalep, 214 at the junction of the Kaiya and Porgera rivers, 259 at Paiam, 609 at Yuyan, and 644 at Tipinini (Wabag PR #2, 1957/58). A census conducted in January 1959 counted 2,126 people, of which 1 in 15 were missing due to pandanus season, according to the patrol officer, which would mean there were about 2,278 people in the valley, closely matching the previous census (Laiagam PR #2 1958/59). While some of the data about pre-mining populations in the upper Porgera are contradictory, the censuses indicate that the majority of the population (62 to 75 percent) lived in the lower to midportions of the andakama.[8] Several of the patrol reports mention that people refused to live in the lowest reaches of the Porgera valley for fear of spirits and the presence of malaria. By the 1980 census 51.8 percent of the people still lived in the lower Porgera region.[9] At the time of the opening of the mine in 1990, only 37.5 percent of the population still lived in the lower Porgera. By 2000 the lower Porgera population was only 22.6 percent of the total population in the valley (see table 6.2).

Increases in population within the Special Mining Lease (SML) demonstrate that much of the population increase is driven by the incorporation of outsiders. Porgera Joint Venture conducted censuses in 1994 and 1999 within the SML (data from Johnson 2012: 7). In 1994, out of a total of 5,257 people living in the SML villages, only 1,104 (21.0 percent) were not from one of the landowning clans. By 1999 there were 8,710 people counted

Table 6.2 Population change in Upper and Lower Porgera Valley

	1957	%	1980	%	1990	%	2000	%
Upper Porgera	767	38.0	2,382	48.2	6,263	62.5	17,650	77.4
Lower Porgera	1,253	62.0	2,564	51.8	3,758	37.5	5,159	22.6
Total	2,020		4,946		10,021		22,809	

in the SML. Of these, 4,094 (47.0 percent) were not landowners. Overall, while the population of landowners increased by only 11.2 percent from 1994 to 1999, the population of outsiders increased by an incredible 270.8 percent in this same period. The exploding population is contributing to substantial environmental changes in the upper Porgera region.

Satellite Imagery and Deforestation

Aerial photos and satellite imagery allow us to track the movements of people in the valley beginning in the early 1970s via forest clearance, although by this time there was a thriving indigenous alluvial mining scene already in place. As would be expected, the upper Porgera shows considerable impacts in terms of forest clearance and settlement over time. Satellite data from Porgera are available from 1972 to the present; however, obtaining cloud-free images of the PNG highlands is difficult. As a result I chose images documenting decade-scale changes and analyzed them for land cover change in ArcGIS. I selected satellite imagery from the years 1972, 1982, 1991, 2002, and 2013 to approximate land cover changes occurring over a roughly ten-year time span (see Map 6.2). Since seasonality is fairly minor in Porgera, acquiring images from the same month was not a great concern, although four of the five images were acquired between July and September.[10] The 1972 data compilation was augmented with aerial photos taken over Porgera in 1973 by the Australian Survey Corps. Satellite images were preprocessed in ArcGIS through histogram equalization and cubic convolution resampling. Supervised classifications using maximum likelihood classification were run on the images to obtain three land classes: forest, gardens, and bare land or disturbed areas. Images were postprocessed to remove small clouds and to cross-check land cover classes against other satellite images near the same time period. Finally, images were clipped with a mask matching the southern Porgera River watershed boundaries (an area equaling 41,972 hectares).

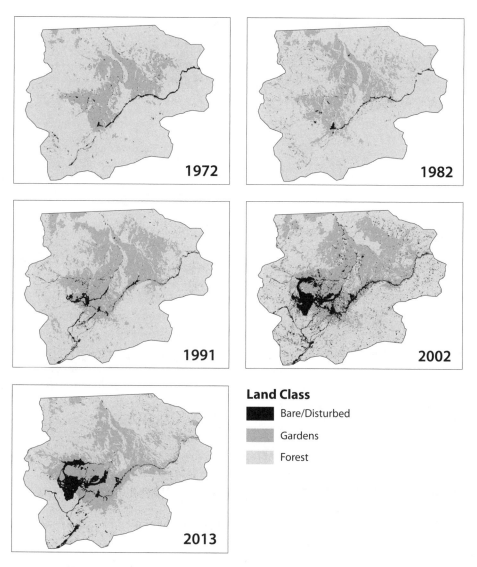

Map 6.2. Land cover changes in southern Porgera watershed, 1972–2013.

Overall the most significant change in land cover from 1972 to 2013 has been the loss of rain forest and the expansion of bare or disturbed land and gardens up until 2002, followed by expansion of forest and declines in the other two land classes more recently. In 1972, 80.3 percent of the southern Porgera valley was covered in rain forest. By 2002 some 8,500 hectares of forest had been converted to gardens, leaving only 60.2 percent of the southern watershed forested (see table 6.3). Gardens

Table 6.3 Hectares and percentage of land cover by land class

	Bare/disturbed	Gardens	Forest
1972	478.4 (1.1%)	7,781.6 (18.5%)	33,669.0 (80.3%)
1982	354.3 (0.8%)	8,252.5 (19.7%)	33,361.3 (79.5%)
1991	854.4 (2.0%)	12,027.9 (28.7%)	29,085.7 (69.3%)
2002	3,178.5 (7.6%)	13,534.8 (32.2%)	25,253.0 (60.2%)
2013	2,052.2 (4.9%)	10,498.0 (25.0%)	29,417.2 (70.1%)

increased over this same time period by almost 6,000 hectares, from 18.5 percent of the total area to 32.2 percent. Bare land and disturbed areas went from 1.1 to 7.6 percent of total land area.

One of the remarkable findings in table 6.3 is the reforestation between 2002 and 2013, despite the population more than doubling (using PJV estimates for 2013). Forest cover increased by almost 10 percent of the total area. In the preceding decade (1991–2002) the same doubling of population resulted in a loss of approximately 4,000 hectares of forest. Figure 6.5 demonstrates this graphically. There are several factors related to reforestation. First, Porgerans have restricted the ability of outsiders to cut forest for firewood and building materials. Second, increased cash income leads to less reliance on subsistence gardens as people purchase rice and other staples to replace or supplement sweet potato (i.e., decreased total area for gardens from 2002 to 2013). Third, conflicts, especially in the southeastern Porgera valley, have led to the abandonment of entire communities as people have migrated elsewhere to avoid the fighting. Indeed the rate of reforestation between eastern Porgera and western Porgera is almost twice as much in the eastern part of the valley from 2002 to 2013 (21.7 versus 12.4 percent).

Linking forest clearance to the indigenous ecological zones recognized by Porgerans (see figure 3.4) demonstrates that changes in land cover are not uniform across the region in the southern portion of the Porgera valley. In the lowest elevation zone, below 1,600 meters (wapi), forest cover varied by only approximately 200 hectares over forty years. In the middle elevation zone, from 1,600 m to 2,200 meters (andakama), forest cover declined to 28.4 percent by 2002 but increased to just over 50 percent by 2013. In the highest elevation zone (aiyandaka), forest decreased from 93.9 percent in 1972 to 73.0 percent in 2002 and increased

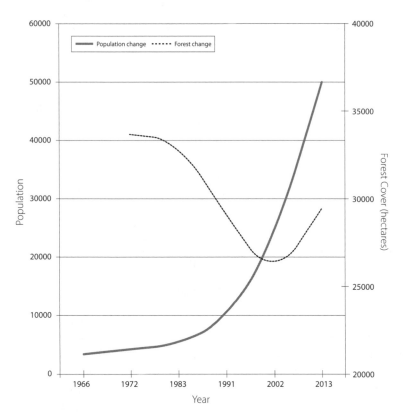

Figure 6.5. Changes in population (left y axis) and forest cover (right y axis) over time in Porgera.

slightly by 2013 to 78.2 percent (see table 6.4). Again an interesting finding is that areas dedicated to gardens across all three zones decreased from a high of over 13,000 hectares to just over 10,000 hectares by 2013 despite a doubling of the population, indicating that people are pursuing two strategies in the face of increasing population: intensifying their agricultural efforts (see also McAlpine and Freyne 2001) and increasing the amount of purchased food items.

Considering that people rarely garden above 2,400 meters in Porgera, forest loss in the highest elevation zone is actually more substantial than in any other elevation zone. Of a total of 28,876 hectares of total land area in the highest elevation zone, 23,453 are above 2,400 meters. This leaves only 5,423 hectares of land available in the aiyandaka for agriculture. As shown in table 6.4, Porgerans exceeded this number in 2002, and many gardens adjacent to the mine and the highway were cut as high as 2,600 meters in the highest elevation zone. Overall the

Table 6.4 Hectares and percentage of land cover by land class and elevation zone

	Below 1,600 m		1,600–2,200 m		Above 2,200 m	
	Forest	Garden	Forest	Garden	Forest	Garden
1972	1,196.2 (67.7%)	547.2 (31.0%)	5,408.7 (47.7%)	5,729.5 (50.5%)	27,094.1 (93.9%)	1,504.9 (5.2%)
1982	1,117.3 (63.2%)	643.3 (36.4%)	6,137.8 (54.1%)	5,093.0 (44.9%)	26,106.2 (90.4%)	2,516.2 (8.7%)
1991	1,106.4 (62.6%)	643.1 (36.4%)	3,704.4 (32.7%)	7,427.6 (65.5%)	24,274.8 (84.1%)	3,957.2 (13.7%)
2002	969.9 (54.9%)	738.9 (41.8%)	3,217.6 (28.4%)	7,245.4 (63.9%)	21,065.5 (73.0%)	5,550.5 (19.2%)
2013	1,096.4 (62.0%)	663.2 (37.5%)	5,737.5 (50.6%)	5,040.7 (44.5%)	22,583.3 (78.2%)	4,794.1 (16.6%)

pattern of deforestation in Porgera shows that the heaviest impact occurs in the highest elevation zone adjacent to the highway. Analyzing deforestation within a 1-kilometer buffer around the highway demonstrates that in 1972, 60.7 percent of all lands in the buffer were forest; by 2013 this had decreased to 44.1 percent. Moreover, with increased road building due to mining development, the total hectares of land included within a 1-kilometer buffer zone rose from 5,138 in 1972 to 12,718 by 2013 (map 6.3).

The change in land cover inside a 2-kilometer buffer around just the highway—thereby keeping the area constant—shows forest cover remaining constant from 1972 to 1982. Following the opening of the mine in 1990, the forest cover declined from approximately 67 percent in the preceding decades to 50.4 percent by 1991. A subsequent decline occurred the following decade, with forest cover decreasing to 39.2 percent by 2002. An increase of approximately 300 hectares of forest by 2013 increased forest cover in the 2-kilometer buffer to 42.7 percent. Figure 6.6 shows the percentage of land cover in a constant 2-kilometer buffer (10,336 hectares) every decade from 1972 to 2013. Again this supports the findings that Porgerans have been able to limit the levels of deforestation even with increasing immigration into the valley and that later migrants (mostly non-Porgerans) cause less deforestation than initial migrants (mostly internal, Porgeran migrants). However, forest degradation is a complex issue in Porgera and is tied to both social and ecological factors.

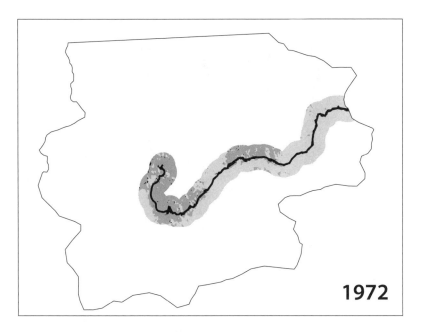

1972

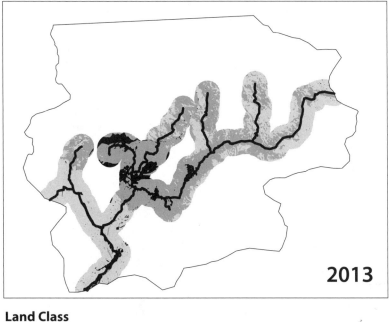

2013

Land Class

Bare/Disturbed   Forest

Gardens   —— Roads

Map 6.3. Land cover change within a 1-kilometer buffer around roads, 1972 and 2013.

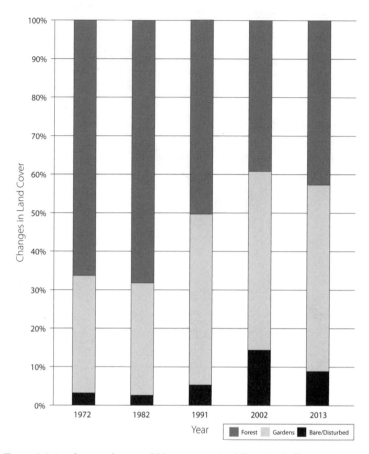

Figure 6.6. Land cover change within a constant 2-kilometer buffer around roads.

Changes in Species Inventories in Secondary Forests

There is strong evidence globally that road building in tropical rain forests results in deforestation (Curran et al. 2004; Laurance et al. 2009; Oates 1999). However, as my data indicate, while roads serve as magnets for immigration they also potentially prevent greater deforestation in hinterlands—the lower rain forest, in the case of Porgera. Later migrants also cause less forest loss than initial colonists to an area (Laurance et al. 2009). The critical finding is that mining development has led to greater deforestation in Porgera, but it is not uniform across the region. There are spatial and temporal pulses in forest degradation. I inventoried the different species that are found in primary and secondary forests at the different elevation zones. I also linked the ecology of forest succession to forest management practices of Porgerans to explore the variable effects of deforestation at different altitudes. While anthropogenically planted casuarina are perhaps the most domestically important tree species, naturally occurring trees in secondary and primary forests are also a critical resource for when people need more wood than can be supplied by casuarinas, which is fairly often. As will be seen, altitude has a significant impact on which trees make up secondary forests, thereby hampering the ability of Porgerans to obtain adequate wood for building and cooking in the higher altitude zone.

I conducted a survey of tree species by land use type and elevation zone in the eastern part of the Porgera valley around the Kairik airstrip (5°26' S, 143°10' E). This was land that was held by the Bipe clan, among whom I was living in 2006. I chose this site due to its proximity to the Enga Highway and because there were several "camps" of mixed Ipili and Enga living nearby who were using trees for house building and firewood, allowing me to explore population pressure on local resources and on tree tenure and land tenure dynamics between immigrant and host communities and individuals. I sampled twenty-two transects measuring 50 by 10 meters (1.10 hectare in total) that were established in both primary and secondary forest at each of the three elevation zones, with the exception of primary forest in the upper elevation zone, as there is no longer any primary forest in the region around the Kairik airstrip. What species should have been in primary forest in the upper elevation zone I reconstructed from oral history and data from other sites in the PNG highlands at a similar altitude.

In the lower elevation zone, I sampled three areas (1,500 m²) in primary forest and four areas (2,000 m²) in secondary forest. In the mid-elevation zone, I sampled five areas (each 2,500 m²) in both primary forest and secondary forest. In the upper elevation zone, five areas (2,500 m²) were sampled in the secondary forest only. The secondary forest sites ranged in age from eight to twelve years since gardens had last been planted in them. Approximate ages of the secondary forest sites were cross-checked with different informants. Initial locations of all twenty-two plots were chosen by stratified sampling, and direction of plot transect was randomly generated, with the exception of ensuring that cliffs and waterways were avoided. All trees with a diameter at breast height (DBH) greater than or equal to 10 centimeters were recorded (Ipili name taken) and measured.[11] Interviews and focus groups conducted after the sampling determined the various uses for every species (n = 110) encountered.

There is a wide variety in the composition of forests in the highlands, although ecologists have identified highland forests by "nodal" species that predominate in certain forests—oaks (*Castanopsis acuminatissima* and *Lithocarpus* sp.) in the lower montane forests and beeches (*Nothofagus* sp.) in the midmontane (Johns 1982; Paijmans 1976). Porgera fits this characterization well, according to my data; oaks make up 40 percent of all species in the lower elevation zone, and oaks and beeches account for 40 percent of all species (11 percent beeches, 29 percent oaks) in the midelevation zone, which corresponds to the boundary between lower montane and midmontane rain forest determined by ecologists. In terms of importance value, *Castanopsis* and *Nothofagus* account for over two-thirds of this value forestwide. Across all of the sampled sites, I recorded 934 individual trees representing 110 species. Four or fewer individuals represented sixty species (54 percent of the total). Twenty-two species (20 percent of the total) were represented by only one individual. Only eleven species (10 percent of the total) were represented by more than twenty individuals. In summary, Porgera's forests have high species diversity but low evenness, with most plots dominated by a few species, what Balée (2006) calls "oligarchic forests" (see table 6.5). For instance, at the family level, Fagaceae (oaks) is the most prominent family, constituting 29 percent of all trees. This is followed by Pandanaceae (pandanus) at 11 percent, Euphorbiaceae (spurges) at 7 percent, and Cyatheaceae (tree ferns) at 6 percent.

As would be expected, however, the abundance and distribution of trees in Porgera vary by elevation and forest type (primary versus secondary). Species abundance can be measured in two ways: by the number of individuals of a species per unit of land or by the amount of wood available for use by a species per unit of land (Chapman and Chapman 2004; Struhsaker 1997). In terms of the amount of wood available for use, foresters use a measure called basal area, which is expressed in square meters per hectare. A final measure that ecologists use to compare the composition of different species in different-size forest plots is importance value (Balée and Campbell 1990; Krebs 1989). This is a relative ranking of species based on three criteria: (1) species frequency (how many plots a species occurs in divided by total number of plots), (2) species density (number of individuals per hectare), and (3) species dominance (basal area). Values based on these three criteria are relativized (absolute values are expressed as a percentage of the sum of the absolute values) and are ranked on a scale of 0 to 300.[12] As my concern in this chapter is to understand which species of trees are economically important, I focus on importance values to highlight which species in each zone and forest type are frequent, dense, and dominant. However, ecological "importance" value does not always correlate with cultural and economic importance from the perspective of local users. In terms of traditional ecological diversity measures, primary forests in the mid-elevation zone have the lowest diversity both in richness and in other diversity measures (table 6.5). Secondary forests in the midelevation zone, however, have the highest diversity.[13] These diversity measures indicate that midelevation forests are well suited for human occupation because in their primary state they are low in diversity but high in important trees, while in their disturbed, secondary state they provide higher diversity than either upper or lower elevation forests.

In the lowest elevation zone, oaks (*Castanopsis* and *Lithocarpus*) are indeed the most numerous species in both primary and secondary forests, as indicated in the ecological literature. In the middle elevation zone, *Castanopsis* oaks are the most dominant species, followed by *Nothofagus* beeches. In the upper elevation zone, *Nothofagus* beeches are the dominant species in primary forests based on informants' information, other studies in highlands PNG (Ash 1982; Manner 1981; Paijmans 1976; van Valkenberg and Ketner 1994), and my own work in upper elevation forests in Porgera outside of the area sampled in 2006. Of extreme interest is that in upper elevation secondary forests, no individuals of *Nothofagus* greater than or equal to 10 centimeters DBH were found in any of the plots, indicating that these trees are either not regenerating in upper elevation secondary forests or are doing so at a very limited spatial scale. The most numerous species were tree ferns (*Cyathea* sp.), followed by a soft-wooded pioneer species (*Homalanthus nervosus*) and the tapuna variety of nut pandanus trees. Table 6.6 summarizes the data on importance values in the sampled transects by elevation zone and forest type.

While the data in table 6.6 show the most important species in each type of forest and elevation, not all trees are economically as important as others, especially in terms of hardwoods used for firewood, fencing, and building materials. Besides which species are present in each zone and forest type, another consideration is that the actual amount of wood per forest and elevation is highly variable as well. These results are summarized in table 6.7.

As seen in table 6.7, basal area is highest in the primary forests in the midelevation zone. This is a function of the size of *Nothofagus* trees found in this zone, many of which had a DBH of over 100 centimeters. The difference between basal areas in the secondary forest in this zone highlights the absence of exceptionally large trees. In the lowest elevation zone, there are no southern beeches. *Castanopsis* oaks predominate and are much smaller in diameter in general than southern beeches; hence

Table 6.5 Diversity indices by elevation zone

	Area (in ha)	Richness (# of species)	Abundance (# of individuals)	Mean diameter at breast height	Shannon-Wiener Index (H')	Simpson's Reciprocal (1/D)	# of species = 50% or more of Importance Value
Upper primary	–	–	–	–	–	–	–
Upper secondary	0.25	47	216	15.8	3.33	19.6	9
Middle primary	0.25	29	148	33.3	2.35	4.9	2
Middle secondary	0.25	65	242	17.2	3.53	18.9	11
Lower primary	0.20	34	160	24.1	2.94	11.8	5
Lower secondary	0.15	31	168	18.8	2.69	9.0	5
Totals	1.1	110	934				2

$H' = (-1)$ *the sum of $P_i * ln(P_i)$ where Pi is the proportion of individuals of one species calculated against the total number of individuals in the sample and ln is the natural logarithm.

$1/D = 1$ divided by (the sum of $(n*(n-1)))/(N*(N-1))$ where n is the number of individuals of one species and N is the total number of individuals in the sample.

Table 6.6 Trees in each zone and forest type with an importance value greater than 20

	Species	Ipili name	Importance value
Upper secondary	*Cyathea* sp.	tambu	30.4
	Homalanthus nervosus	ambopaiya	27.4
	Pandanus julienetti	tapuna	25.0
Middle primary	*Castanopsis acuminatissima*	pai	100.3
	Nothofagus sp.	tato	50.1
	Syzygium sp.	pipi	26.3
Middle secondary	*Castanopsis acuminatissima*	pai	52.4
	Nothofagus sp.	tato	25.7
Lower primary	*Castanopsis acuminatissima*	pai	68.4
	Araucaria cunninghamii	taiya	32.9
	Lithocarpus sp.	yomboto	22.2
Lower secondary	*Castanopsis acuminatissima*	pai	64.1
	Lithocarpus sp.	yomboto	38.9

Table 6.7 Basal area per forest type and elevation zone

	Total basal area	Basal area of hardwood species	Hardwoods as % of total
High-elevation secondary forest	17.4	9.6	55.2
Midelevation primary forest	78.9	78.6	99.7
Midelevation secondary forest	26.3	22.1	84.2
Low-elevation primary forest	54.8	54.6	99.6
Low-elevation secondary forest	37.3	36.3	97.3

the lower secondary forest does not exhibit such a dramatic difference in change in stand basal area as the midelevation zone. The highest elevation zone shows the lowest amounts of useful wood per hectare, despite this being the area with the highest population in Porgera. No hardwood species were listed as being economically important in terms of firewood, house building, or fence construction. However, this zone does provide a greater abundance of resources, such as pandanus nuts, pandanus leaves that are used for roofing materials, and tree fern fronds, which are cooked in earth ovens and eaten. Basal area of hardwood species would be higher in high-elevation primary forests than secondary forests (if there had been any left in the study area). In 1999 I observed the clearance of high-elevation primary forest throughout the eastern Porgera valley. Many of the trees that were cut down were huge, buttressed *Nothofagus* trees. Figure 6.7 shows a portable sawmill in operation in 1999 in the high-elevation primary forest just to the east of the 2006 study site; the stumps and planks are from *Nothofagus* trees.

To understand which trees besides casuarina were of obvious economic importance, I conducted freelisting exercises where I had people rank tree species according to the economic category that they were used for. The categories that my research assistants and I decided we should use were firewood, house construction, and fence building. Table 6.8 is a composite ranking of the participants who freelisted and ranked the economically important species.

Of all of the species that were ranked as being important, only two appear in all three categories: *pai* (*Castanopsis acuminatissima*) and *lepa* (*Lithocarpus* sp.), both oaks in the Fagaceae family. Moreover only eleven species in total were used to rank economic importance in all three categories. Whereas *Castanopsis* are the most abundant tree other than *Casuarina oligodon*, the lepa oaks are rather uncommon. For the purposes of abundance, I used the encounter rate across the 1.1-hectare plots where forty or more individuals per species was "High," twenty to thirty-nine was "Medium," and fewer than twenty was "Low." Many of the trees that appear in these rankings are low in abundance; it is unknown if this is a factor of being selectively harvested due to their economic importance, or if they are uncommon ecologically. Using importance values to rank the eleven species from table 6.8 gives table 6.9.

The data in table 6.9 support the abundance rankings from table 6.8. The top four species, *yawale*, *pai*, *tato*, and *yomboto*, are the only ecologically abundant tree species that are also economically important. Yawale (*Casuarina oligodon*) were not used in the calculation of importance values as the numbers of these trees from the sample plots do

Figure 6.7. Portable sawmill in the upper rain forest.

not accurately reflect their actual numbers in the region since people plant them in gardens and around their houses.

The combined results of the ecological and socioeconomic data point to only three species of trees (pai, tato, and yomboto), in addition to one species of anthropogenically planted tree, casuarina (yawale), as occurring in sufficient enough abundance (in terms of both numbers and basal area) for human exploitation for use as firewood, fencing, or building materials. Only one of these three species, tato (*Nothofagus* sp.), occurs in high-elevation forests. However, this species was not found in any of the plots in upper elevation secondary forests, which indicates that overharvesting of parent species or some other factor is limiting their ability to colonize secondary forest settings at high elevations. Midelevation forests contain both pai (*Castanopsis acuminatissima*) and tato (*Nothofagus* sp.) in large numbers. Based on importance values (see table 6.6), primary forests are twice as important as secondary forests for these two species at midelevations. Yomboto (*Lithocarpus* sp.) are a very marginal tree in midelevation forests (importance value of 1.67 on a scale of 0 to 300). *Nothofagus* are not found in low-elevation forests. The abundance of pai can be explained by the fact that it flourishes with some human disturbance, due to its readily coppicing (Paijmans 1976: 89). This is similar to other research on human disturbances of *Castanopsis acuminatissima* around Wau in Morobe Province (van Valkenberg and

Table 6.8 Composite freelist ranking of economically important trees

	Firewood			Houses			Fences		
Rank	Species	Ipili Name	Abundance	Species	Ipili name	Abundance	Species	Ipili name	Abundance
1	Casuarina oligodon	yawale	High	Castanopsis acuminatissima	pai	High	Dodonaea viscosa	lokai	Low
2	Casuarina papuana	yopiyopi	Low	Lithocarpus sp.	lepa	Low	Lithocarpus sp.	yomboto	High
3	Lithocarpus sp.	lepa	Low	Nothofagus sp.	tato	High	Castanopsis acuminatissima	pai	High
4	Castanopsis acuminatissima	pai	High	Compositae	wapeyango	Low	Casuarina oligodon	yawale	High
5	Cunoniaceae	kaina	Low				Unknown	longai	Low
6	Dodonaea viscosa	lokai	Low				Lithocarpus sp.	lepa	Low
7							Unknown	kea	Low

Table 6.9 Number of individual trees per species and importance value forestwide, based on all plots

Species	Ipili name	# of individuals	Importance value
Casuarina oligodon	yawale	n/a	›148.61
Castanopsis acuminatissima	pai	183	148.61
Nothofagus sp.	tato	42	53.02
Lithocarpus sp.	yomboto	46	34.94
Lithocarpus sp.	lepa	12	15.15
Unknown	longai	8	14.08
Cunoniaceae	kaina	6	12.00
Casuarina papuana	yopiyopi	4	10.34
Dodonaea viscosa	lokai	6	9.65
Unknown	kea	1	2.19
Compositae	wapeyango	0	0.00

Ketner 1994). Van Valkenberg and Ketner also looked at the response of *Nothofagus* to disturbance (anthropogenic, landslides, fire, storms, etc.) and concluded that both *Castanopsis* and *Nothofagus* need disturbance to regenerate, but also need mature species to produce seeds: "At various places ... the forest is no longer able to recover to maturity and if intensity and frequency of human activities increase, species such as *Nothofagus pullei* and *Castanopsis* will also disappear because of lack of seed" (53).

These results have great significance given the settlement patterns associated with mining development at Porgera. Two major findings highlight the issues surrounding contemporary concerns over access to forest resources in Porgera: in terms of economic importance, secondary forests in high-altitude zones are essentially "useless" for local livelihood purposes, and despite increased population and resource use in the middle altitude zones, economically important tree species remain robust and fairly numerous. Given that the greatest population increases have occurred in the high-altitude zones of Porgera, the lack of useful hardwood species for firewood and construction in these areas exacerbates livelihood vulnerabilities for the vast majority of Porgerans who are not part of the mining benefit streams or are outsiders who are unable to access forest resources in other parts of the valley. Contrary to linear assumptions of population increase and resource decline, the data suggest that midelevation *Castanopsis*-dominated forests are more resilient to population increases than high-elevation *Nothofagus*-dominated forests. In essence, if the colonial government had built the vehicle road through the middle elevation zones where Porgerans were initially living, there would likely be decreased conflicts over access to forest resources.

Mining development has thus exacerbated ecological vulnerabilities in complex ways by encouraging settlement in the high-altitude aiyandaka to live adjacent to the highway. Social factors also contribute to the tensions surrounding mining development and highway life in Porgera.

social disruptions & ongoing impacts they have.

CHAPTER 7

Social Dislocations

Work, Antiwork, and Highway Life

Interview with Cowboy Kuraia about the World Bank

Cowboy: Jerry, have you ever heard of the World Bank?

Jerry: Yes, why?

Cowboy: Everyone around here says that before Placer [Placer Dome, the initial company involved in running Porgera Joint Venture's mining operation] was a mining company, they got their start by forming a *raskol* [criminal] gang. I heard that they broke into the World Bank and stole a bunch of money and then came here and started a mining company.

Jerry: Really? Who told you that?

Cowboy: That's just what everyone around here says.

This chapter explores the social vulnerabilities of life in the upper elevation zone, the aiyandaka, as people have shifted their settlement patterns to be near the Enga Highway. While there are considerable ecological vulnerabilities introduced by such an internal migration, there are also social pressures that often find their outlet in violent conflict. Central to these social pressures are concerns about what constitutes a quality life in an era of resource development. What are wealth, work, and money? What kinds of new social forms do they entail for the people of Porgera?

Q 1

Work and "Free" Money

Cowboy's analysis of the relationship between international finance, multinational corporations, and resource extraction is illustrative of the way many Porgerans feel about gold mining and the wealth it generates. Unlike customary forms of wealth, wealth that comes from gold is not contained within the usual social and economic boundaries (Dundon 2002). Money that comes from mining is often considered to be useful only in terms of "eating" (in Tok Pisin, *kaikai mani bilong gol*, i.e., immediate, conspicuous consumption). In 1989 Lita Muyaka heard that his clan had become embroiled in a fight while up at Mt. Kare looking for gold: "I found Jonah [a clan brother] and we rented a helicopter using gold nuggets to pay the pilot. We had the helicopter fly us back to Tipinini and we got our bows and arrows. Then we had the helicopter pilot drop us off behind the enemy so we could shoot at them from behind and surprise them." Several of the men listening to Lita tell me this story howled with laughter, but then Lita wistfully reminded them that their clan could have been rich from Mt. Kare if it hadn't been for the compensation payments that had to be paid for the six men's lives that were taken in this episode of fighting. In another case Samson Kaiyama said, "When I heard that they had found gold at Mt. Kare I went up there to get some. I filled up two coffee jars and got K17,000 [about $20,000] for just two jars of gold. After I got my money I flew to Port Moresby and just ate money. I partied and gave a bunch of it away." Cowboy told me that during the Mt. Kare gold rush he would always buy a Coca-Cola with a K20 note and tell the storekeeper to keep the change (cf. Vail 1995). People do, of course, channel mining money into business ventures, bridewealth exchange, and bank accounts, but wealth that comes from mining carries a considerable taint that wealth earned in other ways does not. Samson noted that he bought a used fourteen-seat Toyota van to start a business but wrecked it shortly after he bought it. He shrugged off the loss by stating that he had bought the van with *mani nating* ("nothing money," i.e., worthless money). Among the Huli who mined gold at Mt. Kare, Jeffrey Clark (1993: 744) argues, money derived from gold was "destined to be 'wasted,' used in consumption only and not in *bisnis* [Tok Pisin: business]."

A significant part of the attitudes about gold comes from the relationship between production and wealth. Mining wealth is "wealth without work" (Reed 2009: 3), and its accrual denies the important role of work that underpins the Porgeran social-ecological system. In this sense many Porgerans see the wealth that comes from mining development as a form of "antiproduction." Drawing on an analogy with Marx's (1990) concept of the commodity,[1] customary items of production in Porgera are examples of "congealed" or "objectified" labor. Money from mining is "free," without labor. Hence it has no local value, no form of indigenous concept of labor tied to its production (Munn 1986). Porgerans' critique of mining wealth is more than just economic (Clark 1993; Kalb 2013) as it undermines the social, spiritual, ecological, political, and economic worlds. As such, attempts to extort money derived from mining-related payouts do not generate the same sense of social opprobrium that stealing pigs or garden produce would.

Porgerans distinguish three subcategories of work (piape in Ipili). The first and most basic of these is unmarked and captures all of the customary work that people do on a daily basis: working in their gardens, hauling and chopping firewood, cooking, caring for one another, tending animals, and so forth. The two other kinds of work, wok mani and wok lotu, "money work" and "church work" in Tok Pisin, respectively, are marked categories and often oppositional to one another. Wok mani is any kind of work that generates an income. It usually refers to salaried or regular employment (of which there is very little in Porgera) but is also used to describe the labor involved in growing cash crops and selling them in the local market. Being employed at the Porgera mine is a form of wok mani; however, receiving a mining royalty check for being a landowner is not. For many men—I never heard a woman use the term—wok lotu is used to define an oppositional status to wok mani. The term has a multivalent usage; on one hand it is used proudly, almost defiantly to highlight disdain with the contemporary monetized culture of Porgera. Working with God, as a pastor, lay preacher, or elder, is a higher calling than money-based employment. On the other hand, wok lotu is also used defensively by nonemployed men when they are asked about their current employment status. In interviews, when asked if they had a job, many men would defer to customary work by saying things like, "Mi asples man tasol, wok long gaden, brukim faiwod. Mi wok lotu, em tasol" (I'm just a village man, working in the garden, chopping firewood. I just do church work, that's about it).

Many people see wok mani as a desirable state of affairs. Countless times I was asked to either bring or convince a "company" to come set up business in Porgera. Matokas Elata pointed out to me, "We don't need another mining company. All mining companies care about is taking our gold. We need a real company, one that will give us good jobs and stay here for a long time." Pursuing a wok mani lifestyle presents its own set of challenges, however. Many men explained that getting a job at the mine, while initially exciting, soon lost its luster. The ten-day-on, four-day-off, twelve-hour shifts were grueling for a people unaccustomed to following clock and calendar time schedules. On paydays nonemployed friends and kin would show up from near and far demanding money for sick relatives, children's school fees, and other needs. As one recent work leaver noted, "It's easier to just live free in the village. Free wood to build a house and make fires, free land to garden on, free water—no one bothers you. Wok mani is too much of a problem." As the mine refuses to house local workers in the mining camp, the frequent intergroup hostilities also prevent many workers from crossing enemy clans' lands en route to bus stops along the highway.

People who start their own businesses also face many of these same problems. Carter Joe, a man in his early thirties, discussed the challenges of being a small business owner in Porgera. (Two days before this interview, Carter's father's half-brother, Leapen, and Leapen's son, Koa, had just killed Leapen's uncle [mother's brother] Tundu over some disputed land along the highway that they wanted for a house and business purposes.)

> I'm tired of people following the old ways. Tundu's clan burned down my trade store at Kairik yesterday and I think I lost K12,000—all because Leapen killed Tundu. I had two snooker tables worth K2,500 each and about K7,000 worth of goods. That doesn't even include the cost of the store and all of the little stuff. I often wish I was living in another province, then all of my kin wouldn't come and ask me for money and help all the time. Every time someone needs school fees, or money for the hospital, or there's some kind of trouble, people come and expect me to help out.
>
> I have two trade stores, one in Kairik and one in Kolatika. They have both been robbed a couple of times in the past year, but I haven't gone to court because I'm afraid of standing out as a man who complains about losing money. Only one time have I gone to court over my trade store in Kolatika getting broken into. The people who did it compensated me with two pigs and K2,000, but I won't go to court again because I'm afraid of the retaliations that might result.
>
> The problem with our society is we just don't think about the future. When my brother, James, killed that man at Kukulama [in 1986], he ruined his future life. He also messed up my schooling as I didn't do well on my Grade 10 exams and didn't get accepted into National High [School]. So then James was just sitting around in the village and

I felt sorry for him, so I built a trade store in Kolatika that he could run for me. Koa had a good job at the mine that he managed to get without a Grade 10 degree. Now he's ruined his future and he'll just be stuck in the village [*em bai stap mangi long ples tasol*].

Indeed a few weeks after Leapen and Koa killed Tundu, I saw Koa sitting in front of Carter's trade store in Kolatika selling his PJV mining boots, which were a much coveted item among the many young men who were not employed. I asked Koa why he was selling his boots. He responded, "What do I need boots for? I just live in the forest and garden now. We fought and they [Tundu's clan] would kill me if I ever went back up near the highway again."

That Koa had killed one of his matrilineal relatives over land along the highway elicited little surprise among Porgerans. Just as there are varied perspectives about work and money, there are complex attitudes about highway life. The highway is seen as both the source of contemporary social disruptions in Porgera and a place representing the social and economic development that many Porgerans desire.

Me Ata: "Doing Nothing" and Highway Life

There are phrases in both Ipili and Tok Pisin that capture how people describe life along the highway. In Ipili, people say *me ata* (doing nothing), while in Tok Pisin they use the phrase *raun raun* (wandering around) to discuss their actions, or lack thereof. Better translations of both of these terms would be something like "fucking around," "chilling," "chillaxing," or "hanging out," as they are intentional actions of purposelessness. In a sense they are cultivated forms of antiwork in that one is purposefully in a public setting intentionally doing nothing productive. Boozing, gambling, people watching, and engaging in illicit sex are all nonproductive activities associated with the highway. Those who don't live along the highway characterize those who do as engaged in these activities the majority of their time. When those who don't live along the highway come up to the highway communities, however, they too "do nothing" and raun raun, slipping into a form of cosmopolitanism (Ferguson 1999) that highway life demands. James Ferguson defines cosmopolitanism as "a set of localizing social pressures . . . [that] must inevitably appear, in a localist perspective, as rule-breaking, as a practice of establishing distance from, or even a social rupture with, the 'home folk' who might make claims on an urban worker" (211–12). While Ferguson intended this to describe different

Figure 7.1. Selling betel nut and cigarettes along the highway.

registers of rural and urban behavior in Zambia, in Porgera people raun raun because they have money but intend to "eat" it by drinking and playing cards. In so doing they are able to make manifest their desires for an individualism unconstrained by social obligations. While I was conducting research along the highway (figure 7.1), I would occasionally ask non-highway-dwelling friends if they wanted to join me. They would inevitably reply, "Why would I want to go to the highway? I don't have any money."

The highway is seen as a place of rampant illicit sexuality and immoral behavior, covered by the Tok Pisin phrase *pamuk pasin*, which is also used as a loan term in Ipili conversation. The Tok Pisin phrase *pamuk meri* (*pamuk* woman) is one term for a prostitute or sex worker. The other term for these women in Tok Pisin is *pasindia meri* (passenger women). The term is meant to capture some of the elements of transience that a relationship with such a woman would entail. In this case it would refer to the short-term aspect of sex for money rather than the long-term associations of marriage and bridewealth payments that define conjugal responsibility. According to Wardlow (2006), pasindia meri do not fit the prototype of a sex worker because these women frequently do not exchange sex for money. As one of my male informants said about pasindia women, "They sometimes do it for money, sometimes a smoke or a Coca-Cola. Sometimes they just fuck for free." When I asked why they are called "passenger" women, several people explained that there were no such women prior to mining development. With the coming and going of traffic along the road that services the mine, these women jump into the cars and trucks of men seeking sex.

Pamuk pasin is a popular topic in church sermons. During one service the sermon centered on Revelation 18:2–4. I provide a condensed translation from the Tok Pisin Bible as the sermon was read in Tok Pisin, then translated into Ipili, and there are some slight semantic differences between the Tok Pisin and the English versions:

> Loudly he cried out, "Babylon has fallen [*bagarap pinis*]. Now it has become a place of evil spirits. For all the people have been engaging in pamuk pasin ['fornication' in the King James version] there. They have been drinking the liquor of Babylon. And the kings of the earth have been engaging in pamuk pasin there. And businessmen have grown rich on the evil there and show no shame at their ways."
>
> And I heard the voice of another man, coming from heaven: "My people, you must leave this place and come outside. You must not associate with the sins of this place, for these sinners will surely suffer for their actions."

After the service I stood outside the church and asked a group of people what they understood by the word *Babylon*. They all said the pastor was talking about the highway. My friend Julie told me, "The pastor was saying that Babylon—the highway—is a bad place and, if we go there, we start to flirt [*ai gris* in Tok Pisin, 'eye grease'] and envy and desire for money."

One night in mid-1999, at around 2 a.m., a truck came racing into Kolatika carrying "two men, two prostitutes, and a carton of smuggled liquor," according to the Kolatika residents I spoke with the next morning. The truck was being chased by a police Land Cruiser, but the cargo and the passengers escaped into the forest. By morning an angry crowd of people had gathered around the truck. They decided that they would impound the truck and make the owner pay K1,000 to get it back. I was asked to type a letter stating that "Kolatika is not a place where pamuk pasin will be tolerated. Unlike the highway and the mine site, there is no alcohol and marijuana abuse here, nor are there prostitutes." I typed up the letter, while several men set up a guard around the truck. The people from Kolatika planned to present the letter to the owners or police and demand compensation for having "highway" customs brought to a non-highway community. Around midnight that night we were awakened by the guard crying out, as the owners had attempted to sneak down

Figure 7.2. The reciprocity of bush life: a hamletwide distribution of pork.

in the middle of the night to get their truck back. As I ran out to where the truck was, the moonlight and the truck's headlights cast an eerie pallor over the scene as arguments broke out and men began to wave their machetes in anger. The Kolatika leaders lectured the owners of the truck (who were sml landowners) on the consequences of bringing pamuk pasin there. After paying a K100 fine, the owners were allowed to have their truck back and they left. Everyone present was given K2 or so from the money collected.

Another set of discourses that surround highway life depicts the behaviors of highway dwellers in opposition to those of non-highway dwellers. In local terms, people talk about differences between the *bus* (Tok Pisin; "bush, forest," i.e., areas not located on the highway) and the *main sait* (Tok Pisin; "mine site," i.e., not just the sml area but also the highway communities when used as a contrast to *bus*). The bush/highway discrepancies in behavior are conceptualized in terms of reciprocity and nonreciprocity (figure 7.2). For example, whenever I would eat with people who lived in the bush, I would always thank them as they passed me my sweet potatoes and greens. They would shrug and tell me not to thank them, that "this isn't the main sait, we share down here." In contrast people along the highway *givim baksait* (Tok Pisin; "turn away") and are said to be "selfish," "stingy," and "unwilling to share food and money." Pokali Kelu explained the relationship between reciprocity, development, and social and ecological decline:

Pokali: When young boys get beards and young girls get breasts and periods when they're only eight or so, then you know the ground is close to ending [yu koyo peya]. When everyone has metal-roofed houses and cars and eats good food, then the ground will end. When you think only of money, then the ground will end.

Jerry: So this is what you think about Porgera?

Pokali: Yes, it used to be a good place, but now money has made it bad. The highway and mine site people are no good. The people there are selfish, they don't give you a half cigarette or half Coke, they just smoke and drink it down. The mine site and highway are hungry places too. There's no food there. No one gives you any food either.

Discourses of social order and disorder also permeate narratives related to the bush and the highway. Bush women make good wives because "they listen to their husbands," "work hard," and "stay sitting in the garden." Men say that living in the bush keeps their thoughts "straight" and doesn't make them think about pamuk pasin. As mentioned, the idea of not straying is encoded by the saying "Ke ko, ki epene" (Legs bad, hands good), meaning that people who stay on their land and work are good, while those who venture up to the highway to raun raun invite trouble. The social disorder of the highway is perceived through sexual idioms (cf. Bell 2006 for the Purari Delta) and the fear of sexually transmitted diseases. Many of the women along the highway are said to be sik AIDS meri (women infected with HIV). In 2006 in Kairik (a highway community), I saw a prominent grave site that had been erected adjacent to the main market there. Surrounding the grave were several red cordyline (Cordyline fructicosa) plants, which usually signaled that the man in the grave was murdered and his clan intended to avenge his death. I asked the village councilor who was buried in the grave and why it was so prominently displayed. He said, "There are two sik AIDS meri in there. They died last year and we buried them there like that as a warning to all of our daughters and all of the young women here. We told them that they need to listen to their parents and their elders and to not go sleep around with young men or they would end up like these two women."

From the perspective of the highway dwellers, pamuk pasin is worse in all of the other places along the highway than in their own villages. For them, highway life—while posing some of the risks discussed earlier and in chapter 6—has benefits that outweigh the drudgery of bush life. First among these is the perception that they are sharing in the benefits of development that mining promises. Many people along the highway have illegally tapped into the electrical infrastructure provided by the mine, allowing them to have lights and TVs and other appliances. Having mine workers in the household or village also provides people access to money during pay periods, when wages pulse through the community as debts are repaid and money flows through the ever-present circles of card-playing gamblers. For some, just the excitement of the coming and going of tanker trucks, cargo trucks, public motor vehicles (PMVs), and cars along the highway provides the excitement that life in the bush fails to do. Not surprisingly, many of the young people, male and female, who live in the bush have told me that they find the possibility of highway life alluring.

Factions in Contemporary Porgera

The shift from alluvial to hard-rock mining radically altered social and economic relations in the Porgera Valley (see Biersack 2006a for an in-depth overview). The largest source of alluvial gold lies in the middle reaches of the Porgera River in an area Porgerans call Lower Porgera. Where the mine was to be built, however, was in the upper tributaries of the Porgera River on the lands of different subclans than the ones who owned land in the Lower Porgera. Given the flexible social relations, one might wonder if this was not to be a problem. Wouldn't mining proceeds flow widely through the broad kinship networks? In total, twenty-three subclans in seven different clans owned lands within the boundaries of the Special Mining Lease. With approximately seventy named clans living in the valley, only 10 percent were designated as SML landowners by the conditions of Papua New Guinea's Mining Act. However, the amounts of land held by each subclan were significantly different, which meant that just a few subclans have received the bulk of mining proceeds that come from compensation. Royalties are also paid on the basis of land owned within the SML to agents from the twenty-three subclans, who are then supposed to distribute the money to individuals, although it is widely reported that big men and male heads of the household retain most of the cash (Golub 2014). Therefore, to a large extent, money has not flowed with the same ease and freedom along the kinship networks that people, pigs, and shell wealth formerly did. Moreover, while it was fairly easy to gain access to the alluvial goldfields by activating a distant kinship link, it required a person's labor to extract wealth from the land. In the era of hard-rock mining, wealth comes without work for the SML landowners, and there is less emphasis on sharing access to wealth than there was in the alluvial mining era.

Another significant impact on social relations in Porgera comes from the way that mining waste and tailings were to be disposed. In lieu of traditional tailings dams, the mining consortium convinced the PNG state to allow the riverine dumping of waste rock and tailings, given the high annual rainfall and geological instability of the region. At current rates of processing the mine dumps around 17,000 metric tons of waste in the Porgera and Kaiya (a major tributary of the Porgera) rivers on a *daily* basis. According to interviews with mining officials, initial engineering estimates determined that the mining waste, stored in both "stable" and "erodible" dumps, would in the case of the stable dumps not erode and that the erodible dumps would quickly be washed down the Porgera into the larger rivers of which the Porgera is a tributary. These estimates were wrong in both cases, however, and what has ensued has been the creation of two massive rock "glaciers" slowly moving down the Porgera and Kaiya watersheds. Not only have these waste dumps buried the adjacent alluvial gold beds in the upper Porgera, but increased sedimentation has buried the lower Porgera gold beds under meters of silt and sediment. In 1996 the lower Porgeran alluvial gold miners reached a settlement with the mining company in which they received a one-time cash payment, followed by quarterly payments based on the tonnage of rock waste and tailings dumped into the Porgera River. Many of the miners claim that these payments are paltry compared to the amount of money they made during the alluvial mining era.

The factions that developed between the lower and upper Porgerans also feature in concerns about the ground ending. In the eastern Porgera valley, there was a prophecy that a young Bipe man would go inside Mt. Watukati and kill the pig believed to be living with Kupiane there (see chapter 5). In 1997, when Opis Papo became MP for the Porgera-Lagaip District running on the campaign of being "a sworn enemy of PJV," people interpreted this as evidence that the prophecy was coming true. Opis, a Bipe man, was going to "kill" PJV and the mine. After winning, Opis formed a new national party, the People's Resource Awareness Party. This drove people into the church in 1998, intensifying their fear of the ground ending as local prophecy, national politics, and international development intertwined in what many characterized as a catastrophic impending doom.

The mining benefits flowing to the SML landowners have resulted in what Albert Nita (2001: 167) refers to as the development of a "spending culture," where there is "arrogant spending behaviour on consumer goods with little regard for long term investment." As one community affairs officer said to me after a relocation meeting with some SML land-owners, "I think the meeting went well. The only issue is whether they'll ask for one or two Rolls Royces." Expensive cars, new clothes, luxury food items, and partying trips to Mt. Hagen constitute the bulk of SML landowners' purchases. Porgera has also seen key players within the SML landowning clans becoming big men on a far larger scale than customary political economics allowed. Many of these men spend their days driving around the government station at Porgera in newer model Toyota Land Cruisers with dark-tinted windows (see figure 7.3); nearly all are polygynists, some with as many as twelve wives (Bonnell 1999). One has twenty-five or thirty wives, whom he keeps in a dormitory-like compound adjacent to his house (Johnson 2011: 20). While the SML land-owners and the thousands of non-Ipili outsiders who have flocked to the SML villages and the highway to share in the wealth participate in a world of commodities and Western foodstuffs, a large majority of Porgerans still live a subsistence lifestyle. This makes Porgeran society, as such, highly factionalized. At one extreme are the SML "super big men" (Nita 2001), with their multiple wives, business holdings, and new cars. At the other extreme are the subsistence farmers living far from the hustle and bustle of the highway, whose luxury items are restricted to salt, soap, and cooking oil. Between these extremes are the SML landown-ers who live in relocation houses with electricity yet who may receive very little money from the mining. Also in between these extremes are the non-SML landowners who work at the mine and host a number of kin, affines, and acquaintances who make constant demands on their monetary resources.

There have been some attempts at the government level to reduce the factionalization of Porgeran society. Porgeran development occurs in the shadow of the Panguna mine on Bougainville Island, North Solo-mons Province, which was shut down by landowner protests in 1989 over environmental damages and unequal intergenerational compensation practices (May and Spriggs 1990). As a consequence mining development at Porgera was negotiated in the hopes that money and services would accrue to the greatest possible number of people. Mining officials voiced the opinion that, since the Ipili were "an exchange-based society," min-ing money would move extensively throughout the kinship network. Yet, just as the mining engineers' plans for waste to flow rapidly down the Porgera River failed, so too have the plans from the mine's social development offices failed, for money from mining royalties and com-pensation has not been shared widely. The fact that wealth from hard-rock mining was not making its way into outlying communities was well

Figure 7.3. A health pamphlet put out by the PNG government warning young women of the dangers of catching STDs from *dakglaskaman* (men who drive around in cars with tinted windows).

known by the Porgera District Administration. In 1998 district officials began discussions with a vociferous group of non-SML landowners in the eastern Porgera valley who were upset over "eating the dust of the mine but not eating any of the money," as one leader put it to me. The discussions between the government officials and the residents of Tipinini were to create interest in alternative income-generation schemes to mitigate some of the wealth discrepancies and factionalisms that were developing in the valley. The main focus of this scheme was to start a coffee plantation and nursery, which would serve as a demonstration center and coffee plant source for local farmers.

Despite coffee being one of highland PNG's most important export crops (West 2012), with a gold economy Porgerans never became overly involved in coffee growing. In 1969 colonial patrol officers brought coffee seedlings to Porgera and its neighboring Ipili-speaking valley, Paiela. The subsequent year, however, patrol officers noted an approximately 60 percent rate of mortality for the coffee plants in Paiela (Lagaip PR #2, 1970/71). In a coffee tree census conducted in 1973, there were only 1,105 coffee trees growing in the entire Porgera valley (Porgera PR #7, 1972/73). Over the years many of the trees were abandoned, and in a 1998 census leading up to the coffee development talks, government workers in the Department of Primary Industries (PNG's agricultural department) told me that the trees had grown so large that the workers had to climb adjacent trees just to pick the coffee cherries. Data from the 1998 census indicated that in the Tipinini area (comprising three of Porgera's ten census wards), only seventy-nine households were growing coffee and that there were only 6,950 trees in the area (about eighty-eight trees per household, although the range per household was from 2 to 811). Given that the soils and climate were obviously suitable for coffee growing, the district government launched a two-pronged initiative to promote coffee production: to encourage every household to plant and look after five hundred coffee trees and to develop a 165-hectare for-profit coffee plantation that would serve as a resource center for meeting their first initiative. Farmers would learn how to grow and harvest coffee, and the plantation would also have a nursery that would sell coffee plants to local farmers. The exciting thing for the government was that the plantation would be a partnership between local landowners in Tipinini and the Porgera Development Authority (PDA) that would allow proceeds from mining to directly benefit non-SML landowners. The landowners would supply the labor and the land, and the government would oversee the money that would be provided by the PDA for the plantation development.

While I was living in the Tipinini area conducting dissertation research in 1999, the PDA asked if I would be willing to work as a paid consultant for two months in order to conduct a baseline socioeconomic survey, assess attitudes about cash cropping and coffee growing, and survey opinions about the plantation in the three census wards where the coffee development would be occurring. With my two research assistants, who had already been working with me for several months, we developed a questionnaire and randomly administered it to twenty individual household heads in each of the three different census wards. We also held two gender-segregated focus groups in each of the three census wards to understand differing male and female perspectives on cash cropping and socioeconomic services, and then had the groups rank the services in order of preference. We also held semistructured interviews with some of the current coffee producers, key landowners in the proposed coffee plantation area, and the elected councilors in each of the three wards.

Findings from the research gave us doubts about the long-term viability of the project and its success in reducing factionalism related to mining-related wealth inequalities. The first concern was lack of market access. There was only a footpath to the coffee plantation and the government had no plans or money to build a road. Road maintenance is a constant issue in the highlands of PNG given the constant rain and geological instabilities. In fact, through an instrument called the Tax Credit Scheme, the mine had taken over many of the government services in Enga Province, such as road building and maintenance, in lieu of some of its tax burden being paid to the state. Money from the Tax Credit Scheme had already been spent in the area around the plantation on housing for the elementary school teachers as well as on rain catchment tanks at the school, and the mine was unwilling to put more money into the area for a road. The second concern was the location of the coffee plantation in relation to traditional land holdings. The plantation was to straddle the lands of two clans (technically four subclans, two from each clan), the Pakoa and Yawanakali, who were longtime enemies. A number of deaths that had occurred on both sides within the past decade still had not been compensated. As a consequence there was much distrust between the principal landowners involved in the project. The final concern was the lack of the Porgera Development Authority's willingness to sign any sort of agreement with the landowners about compensation for the land or profit sharing. Over all the years of alluvial and hard-rock mining, Porgerans have come to appreciate formal, written documentation surrounding agreements about land tenure, land use, and appropriate compensation

Figure 7.4. The Maliapaka coffee plantation.

for lands taken out of horticultural production for development purposes (cf. Bell 2009 regarding logging agreements in the Purari Delta).

Despite these concerns the government and the PDA were keen to get the plantation started. Ten contracts were issued to clear one hectare of primary rain forest each. The funders wanted the land cleared quickly, so the holders of the contracts hired between twenty and forty of their group members to help them cut down trees and brush and burn the slash piles. With the contracts priced at K1,000, each person received no more than K50 (about $17) for their efforts. One worker later told me, "There wasn't even time to save the economically valuable trees. It was just cut and burn, cut and burn." The plan was to clear and plant coffee on 15 hectares the first year (despite there being contracts to clear only 10 hectares), and then continue to clear and plant another 10 hectares every year after that so that the 165-hectare plantation would be complete once mining ended. Over the next few weeks, teams of contracted workers (mostly men) planted over twenty-five thousand coffee plants on the 10 hectares (figure 7.4). Excitement in the surrounding hamlets about the coffee development was high. Extra plants were being bought for about U.S. 10 cents apiece and planted around people's households. I bought one hundred plants for each of my two research assistants after they complained that they were missing out on the future wealth from the green gold. I was even starting to get excited about the future of coffee and exhorted my friends to plant five hundred trees around their houses in anticipation of tripling their average household incomes, based on the current price of coffee.

Discussions in the various non-SML hamlets in eastern Porgera centered around forming alliances with the government as a means to gain development. Since the SML landowners had the mine and mining company, non-SML landowners hoped that the government would step in to alleviate wealth inequalities in the valley. Friday Pakena spoke out during one of the coffee plantation focus groups, saying, "Kolatika mas i go develop. Mipela mas stap ananit long gavman, o nogat, mipela bai kamap olsem wel pik" (Kolatika must become developed. We must live underneath the government, for if not, we will become like wild pigs). Other men voiced their support for allying with government institutions. At this same focus group Pore Laikiya argued, "When we used to make salt we always said that you can't let ashes get into the salt or you'll ruin the flavor. The coffee plantation is like salt. Trouble is like ashes. That's why we need some sort of law-and-order committee for the plantation. We can't let trouble disrupt things. If I were a dog, my tail would be wagging because of how excited I am about the coffee plantation."

Unfortunately for people like Friday and Pore who hoped for government assistance with development, profound divisions were forming within government circles from the national to the local level that would reverberate throughout Porgera and diminish support for projects like the coffee development scheme. At the national level, in mid-1999 rumors were circulating through the Enga region that an MP from Enga Province (Kompiam-Ambum District), John Pundari, was going to replace the current prime minister, Bill Skate, after a vote of no confidence. The MP from Porgera, Opis Papo, quickly backed Pundari's candidacy. By July Radio Enga and other news sources that we were picking up on my shortwave radio declared that Pundari would be voted prime minister at any time now. However, on the day of the vote Pundari's party allied with the opposition government, making Sir Mekere Morauta the prime minister. Pundari was given the deputy prime minister position for his last-minute switch. With Opis closely allied with Pundari and the ruling government, people started saying that Opis would be made the mining minister and would make PJV pay for what they had done to Porgera. Puluku Poke, one of the first Porgeran alluvial miners, was overheard in several public settings saying that the white men at PJV were afraid and leaving the valley now that Pundari and Opis were "in charge." To the disappointment of many non-SML landowners, Opis never was appointed the mining minister. In the Sirunki area, east of Porgera, the Enga blocked the highway for several days, looting PJV cargo trucks, when rumors reached them that PJV had paid "millions" to keep Pundari and Opis from power.

In September 1999 factions within the provincial government were exposed during Independence Day celebrations. The Porgera District Administration sponsored a massive day-long celebration with funding provided by PJV and the Porgera Development Authority. As part of the celebrations, a helicopter was sent by PJV to bring Pundari, Opis, and the Enga provincial governor Peter Ipatas to give speeches. When the helicopter landed, only Pundari and Opis disembarked. People carried the two men around the crowd on a flower-bedecked litter to ebullient shouts and cheering (figure 7.5). The first person to speak was the president of the local-level government (LLG), John Pauwe, who said, "The mine is a negative factor in everyone's lives here. People have died, there isn't enough development, and something needs to be done." After Pauwe sat down, the crowd started to get agitated and unruly as several top leaders of PJV were also on the stage. While Pundari was talking, another helicopter landed at the mine's helipad less than a kilometer away. Pundari told the crowd that he and Opis had sent lots of money to Porgera, and if it wasn't being developed it was the fault of the Porgera Development Authority. As the chopper was landing, Pundari was telling everyone how he could have been the prime minister but he was blocked by Peter Ipatas's support of Bill Skate. At about this time a PJV Land Cruiser arrived from the mine with Ipatas. People in the crowd started shouting things like "Why was Ipatas at the mine? Why is he with PJV?" By the time Ipatas stepped up to the microphone, there was a dark tension in the air. As he started to tell his version of the events at Parliament, Pundari and Opis left the stage and headed toward where their helicopter was parked. All of a sudden rocks started flying through the air as people bombarded Ipatas and the PJV dignitaries on the stage. Pandemonium broke out as people ran in all directions. I opened my umbrella to ward off errant rocks and ran to the district office headquarters and crawled underneath the building. People were screaming and blood-spattered individuals staggered past me. After the police fired off fifty or so rounds, everything started to calm down. Ipatas stepped back up to the microphone and told everyone how small they were and how big he was. He said, "I'm the prime minister of Enga Province and can do whatever I like." He then switched to Tok Pisin and said, "Mi gat gutpela senis, mi save draivim gutpela ka, mi save raun long balus. Nau yupela save mekim olsem o nogat?" (I've got fine clothes, I drive a nice car, I fly around in planes. Now do you all do these things or not?). He

Figure 7.5. MPs John Pundari (*waving*) and Opis Papo.

then claimed that he had "built the Enga men's house himself [i.e., the five Enga MP districts]" and that Pundari and Opis were just "little boys" who had recently moved into the house.[2] He told the crowd that he didn't care if they had tried to stone him or not. He pulled a check out of his pocket and said that he was still contributing K5,000 to the celebrations. (A total of K27,500 in prize money was being given out to groups with the best *bilas*—traditional dress—leading one Huli man who was on my PMV to Porgera Station that morning to shake his head in awe and say, "K35,000 [sic] prize money! Porgera really is a money place.")

The critique of mining by John Pauwe, president of the LLG, during the celebrations also highlighted some of the local divisions at play. On one side was the antimining coalition of Opis Papo and the LLG; on the other was the pro-mining Porgera District Administration, the SML landowners, and the Porgera Landowners Association. Epe explained to me how power had shifted with the coming of hard-rock mining in 1990:

> The LLG should be the body with the real power—it handles all of the money at the "grassroots" level. Members of Parliament just sit in Port Moresby and send money to the province, which then distributes it to the LLG. But with the formation of the PDA, the power of the LLG has been eclipsed. Now all the money goes straight to PDA and the LLG gets peanuts. The PDA is all run by outside people. [In the

late 1990s and early 2000s it was directed by a Ghanaian economist named Sam Adams.]

The current president of the LLG is ineffective. When Jonathan Paraia [a key SML landowner leader and one of the people singled out by Nita (2001) as a "super big man"] was LLG president, the relationship with the mine was very strong. The current president has angered the mine, and now PJV no longer meets with the LLG like they used to. This has also cut into our wages [K50/month] since we would get extra meeting pay when we went up to the mine that varied between K20 and K50 per meeting.

The PDA has also taken all of the prestige out of being a councilor. Before, councilors got big names by bringing projects inside their council wards. However, now PDA controls and delegates all of the projects and the councilors get "a bad rap" [Tok Pisin, *kisim nem no gut*, literally "get a bad name"].

Epe's statement highlights some of the ways that being an SML landowner articulates with obtaining key governmental positions. During the late 1990s and early 2000s Kurubu Ipara, a key leader among the landowners, held the district administrator post. In early 2014 Nixon Mangape, another prominent landowner, was the MP for Porgera-Lagaip District and also the PDA chairman. The concentration of power in this way had the potential to overcome some of the factionalism in local government but also highlights the transparency issues that have been leveled against government entities in Porgera (Johnson 2012).

SML landowners have been able to consolidate their wealth and power in other ways as well. As outlined by Glenn Banks (2003: 227), the restructuring of the equity arrangements in 1997 substantially disenfranchised non-SML landowning Porgerans. The original equity controlling company, Porgera-Paiela Yuu Anduane Pty. Ltd., ensured that profits would not only go to SML landowners, PDA, and the LLG councilors but to "other traditional landowners in Porgera." In 1997 the company changed its name to Kupiane Yuu Anduane Pty. Ltd. and restricted shareholding to PDA, which held a 10 percent share, and twenty-three SML landowners representing the twenty-three landowning subclans. In the late 2000s Kurubu Ipara left the Porgera District Administration and was in charge of the entity overseeing the 5 percent equity share held jointly by the Enga provincial government and Kupiane Yuu Anduane Pty. Ltd.

Factionalism between the government, the PDA, and the landowners resulted in the plans for the coffee plantation never being extended beyond the first year of activity in 1999. When I returned in 2006, no new land had been cleared since the initial 10 hectares, and no more coffee trees had been planted due to allegations of misuse of funds directed by the government at the PDA and vice versa. Even more tragically, the killing of Tundu by Leapen and Koa resulted in Leapen's half-brother, Joe—Carter Joe's father—being killed in revenge, which then set into motion repercussions from Joe's clan that eventuated in the destruction of the coffee plantation. In 2003, after Joe was shot in the chest with a shotgun in broad daylight in Carter's trade store, his killer ran off into the forest in the direction of a hamlet adjacent to the coffee plantation. The following day members of Joe's clan and his wife's clan (upon whose lands the killing happened) set out to find his killer. When they arrived in the hamlet next to the coffee plantation, the killer had left, but to punish the people who had harbored him (who were unaware of the murder), they burned down every house in the hamlet and chopped down all of the trees (a common warfare tactic in Porgera). Next Joe's two clans started chopping down coffee trees. Over the next few months a large-scale conflict broke out in the eastern Porgera valley that engulfed all of the stakeholders in the coffee development project. Between 2003 and 2006 seven different conflicts broke out in the eastern Porgera valley that were in some way linked to the initial killing in 1999. In each conflict the coffee trees were a target for destruction as Porgeran warfare seeks to damage the productive resources on which enemy clans are dependent. By the time I arrived in late 2006, depending on whom I talked to, either "all" or "most" of the coffee plantation had been destroyed during fighting.

I was unable to travel to the coffee plantation to assess for myself how much of it, if any, remained. Another conflict had broken out in September of that year, which engulfed the entire eastern Porgera area. This latest conflict was instigated by the same man who had killed Joe in the trade store in 2003. A few days after I arrived in Porgera, I interviewed him—a bittersweet task as both he and Joe had been close friends of mine during my dissertation research. This man, Kangi, had been a cook at the mine and a relative of Tundu. After Leapen and Koa killed Tundu, Kangi was forced to quit his job, as the mining company doesn't provide housing for Porgeran employees, and he would have been forced to travel through enemy lands en route to work. When I knew Kangi in 1999, he had a Tok Pisin name, but now he had changed his name to an Ipili word that means "it is burning." When I asked why he had changed his name, he replied, "After I had to quit and then that trouble happened [his killing of Joe], I knew I wouldn't see money anymore. I was just going back to live in the fucking bush. I didn't want to be reminded of anything modern, so I got rid of my Tok Pisin name." He laughed and said, "I've got a good name now, because now all I do is fight."

The Life Market: The Logic of Warfare as Production

With the demise of the coffee plantation project as an alternative income-generating scheme, the reality of income inequality between people receiving mining benefits and those who weren't was laid bare. Such uneven distribution of wealth generates intensive social pressures that commonly find their relief through warfare. There is a long-standing association between warfare and economics in Porgera expressed in the Ipili phrase *yanda takame*, which means "war is wealth." Marilyn Strathern's (1985: 124) research among the Melpa supports this; she writes that "either exchange or warfare can indeed turn into its alternative." Explaining the term yanda takame to me in 2000, an elder remarked that in the short term fighting is bad because people might die, but in the long term compensations flow back and forth between combatant groups, and eventually, due to the compensations, the groups start to intermarry and live among one another. But by 2006 this association between war and wealth had radically altered. A group of young men from Kangi's clan who were involved in fighting said that now they use the Tok Pisin term *wok long laip maket*, or "working in the life market," to characterize conflict. One of the young men said, "I'm just a bush man. I can't get a job at the mine. I don't have money for school. It's better for me to go to work in the life market. Before, when our ancestors fought, few people died and most were just wounded, then in 2003 we started using guns to fight. Now when you go to fight, you either live or you die. It's one or the other in the life market."

Fighting today, though, is not about generating the long-term benefits of exchange and alliance captured in a concept like "war is wealth," for much of the fighting is generated from dissatisfaction with uneven development and oriented toward short-term monetary gains. For the young men who decide to work in the life market, exchange is no longer an opportunity; instead capturing benefit streams from mining is the new means to become a big man. For them, war is their only option.

The benefit streams entering Porgera have been profound, especially considering that small-scale alluvial mining and horticultural subsistence structured people's livelihoods prior to large-scale development. As of 2010 it was estimated (Johnson 2012) that approximately K6.4 billion in total benefits (royalties, wages, compensation, equity payments, and infrastructure) had been paid out by the Porgera mine. Of this amount, approximately K1.2 billion went to Porgerans, K597 million of it coming as cash from wages, royalties, equity payments, and compensation. In 2009 alone K56 million entered Porgera just from compensation payments, royalties, and dividends (3). PJV does not have good records for the amount of wages that it has paid over the years, but between 1990 and 2009 it is estimated that in total around K183 million in wages were paid to Porgerans (52).

Most of the conflicts today stem from large compensation payments that PJV pays to groups impacted by mining-related development activities, which largely are paid out for loss of land due to mining and other infrastructural developments. The 2006 conflicts that my friend Kangi had instigated were over dissatisfaction with the distribution of compensation money related to a series of landslides that occurred adjacent to the Kairik airstrip. Prior to the development of the airstrip at Kairik in the late 1980s, there was a small ancestral clan lake (ipa ne) called Kaindum at the east end of the present-day airstrip. When PJV was building the airstrip they bulldozed over the lake, forcing the water and the spirits that were in it out of their home. Afterward the area was subject to several small landslides, but PJV never paid compensation because they argued that the landslides were "natural." All of the local landowners, however, knew that the ancestral clan spirits who were angry at being displaced from Kaindum caused the landslides.

One night in October 1999 a powerful storm swept through the Kairik area, causing so much rain to fall that a massive landslide slid across the highway, completely blocking it. An associated downburst completely flattened a 50-by-100-meter area of forest near the landslide. The following day elders said that the flattened forest and the landslide were the result of water from the clan lake falling from the sky and the displaced spirits searching for their former home. The same night of the storm, a group of enterprising young men quickly cleared a path for cars through the landslide but then put up a roadblock charging PJV vehicles K50 to pass through and K20 for all other vehicles. The next day I went up to see the landslide and interview people; I estimated that the young men were earning about K500 per hour from their tariffs. The young men refused to allow PJV near the landslide to clear it, arguing that PJV would first have to assess damages and pay compensation for the people whose homes, gardens, and economically important trees had been damaged by the landslide. Late in the evening that day we heard bursts of automatic gun fire coming from the landslide, indicating that the police had been called in to clear out the young men.

A few days after this event, Isera Pore, the LLG councilor for the local ward where the landslide occurred, asked me to write a letter to PJV detailing the local oral history about Kaindum lake and its significance to the clans in the area. He related that the lake was the home of the yama (clan ancestors) and that in the past people would make offerings of pork to the ancestors there. As discussed in chapter 6, in the late 1980s leaders in the Kairik area urged people to move up to the upper rain forest lands along the highway where the airstrip was going to be built in anticipation of receiving compensation money once construction began. Isera said that when PJV started to build the airstrip, the landowners were "hungry for money" and let PJV destroy the lake, which turned the wrath of the yama against the landowners. He claimed that since PJV owns the Kairik airstrip (*papa bilong epot*) they must pay compensation to the Kairik landowners to placate the yama's anger or else there would be more landslides.

PJV never responded to the councilor's compensation claim. However, the landslides did continue, and after one particularly large one in early 2006, PJV paid K300,000 to the Wapena clan. According to several people, the land that was compensated had long been contested, well before whites ever entered the valley. Apparently in the 1920s or 1930s Kangi's clan, the Piango, were fighting a war against the Watiam clan, which the Piango lost, and most of the Piango left the eastern Porgera valley. The Wapena were allies of the Piango, and during the fighting a Wapena man was killed and his death should have been compensated by both the Watiam and the Piango. However, during the compensations the Piango handed the Wapena a ball of clay wrapped in leaves, which the Wapena interpreted as the Piango handing over their lands in Porgera. The Wapena started using the land as one of their high-altitude hunting areas. Decades went by, and the Piango eventually returned to some of their other lands in Porgera, but as the land along the highway became more coveted during the buildup to mining, Piango clan members tried to move up to the highway for the potential economic opportunities that owning highway land promised. The Wapena prevented them from doing so. In a 1999 land claims case I attended, Muyu pointed out that up until the 1970s, whenever the Piango needed to conduct an ancestral ritual and feed the ancestors at Kaindum lake, no one had ever prevented them from doing so. To Muyu and others, this was evidence that the Wapena still recognized the claims that the Piango had to their former high-altitude lands. One old man stood up during the court and said, "Everyone needs to straighten out the borders between the clans up here. Watiam, Piango, Wapena, Maiyuni, and Bipe all own land up here. We're not the company [PJV]. All the in-married males [*wanakalini*] can stay and have rights and get money too." After his dismissal of the restrictive way that mining benefits were paid out, everyone had a good laugh.

Seven years later people weren't in such high spirits. Of the K300,000 that the Wapena received from PJV, they offered K3,000 to the Piango, which the Piango took as an insult. A few weeks later Kangi killed the LLG councilor of Kairik, who was an in-married male to the Wapena. A month or so after this killing I arrived and moved into Apone Camp along the highway, which was ostensibly on land owned by the Bipe but had several Wapena lineages living there too. The day I arrived in Apone Camp everyone was gathered around a small, corrugated tin shack where 1-kilo bags of rice, cooking oil, and tinned fish were sold. Someone had erected a blue tarp over the front entrance of the trade store, where a cluster of people surrounded an open casket holding the body of a young Wapena man killed the day before. The people under the blue tarp, consisting mostly of close kin, were wailing and crying, lamenting the death of the young man. Several hundred people sat somberly in a large circle around the mourners waiting for the *lakisia* ceremony to begin, in which the clan's warriors would try to communicate with the deceased man's spirit.

After an hour or so, two young men burst forth from the trade store dressed in traditional ritual finery—marsupial fur headbands, long woven bark loincloths, and mother-of-pearl shell necklaces. Their faces were blackened with ashes from burned tires, and they were both carrying shotguns, one a home-made gun, the other a pump-action shotgun obtained in the lucrative illegal gun market (figure 7.6). They ran several times around the casket, calling the young man's name, invoking his spirit to come and indicate whether or not they should avenge his death. Eventually the two men ran into the center of the circle of people and raised the guns to their shoulders to fire them toward the sky. When the firing pins of both guns audibly clicked and yet neither gun fired, everyone started to murmur ominously. The young man with the pump-action shotgun screamed in rage, pumped the shotgun to chamber a new shell, and fired the gun toward the sky. With this attempt, the crack of gunfire echoed off the rain forest–covered cliffs of Mt. Pauwite just to the south of us.

Epe, who was sitting beside me, shook his head, tsk-tsked, as Porgerans frequently do to indicate something astonishing, and said, "Man, this is not good. Both guns misfired." I had never seen a lakisia ceremony before and asked what it meant. "It means that Palia's spirit [the deceased young man] doesn't want to be avenged. Too many people have

Figure 7.6. *Lakisia* ceremony.

been killed in this fight already." Between September and November 2006 nineteen people were killed during this fight. Epe noted that prior to 2003, when guns were first used in fights in Porgera, a big war (*yanda andane* in Ipili) would have seen six people killed. "Now, how will they compensate all the people who are being killed in this fight?" he asked. I pointed out that, with from fifty to eighty pigs needed to compensate every person killed, the two clans would have to find well over a thousand pigs to stop the fight. "It'll never happen," replied Epe, who just tsk-tsked and shook his head.

Two weeks after the lakisia ceremony, Jon, the young man with the home-made shotgun that misfired, was killed by a high-powered rifle shot to his head just a few dozen meters uphill from where Palia was buried. After the guns misfired during the ceremony, the elders had told the young men to stop fighting as Palia's spirit did not want to be avenged. For many of the young men in Porgera, though, warfare is the new economy, the life market (cf. Wiessner and Pupu 2012). In discussing traditional venues of male accomplishment in Hagen, Strathern (1985: 124) writes that the intertwining of war and exchange "can be attributed to men's desire to be seen as effective in a public arena." Excluded from regular mining benefit streams, instead of working in the gardens like their parents, occasionally selling some surplus produce at the local market, these young men offer their bodies in a new market based on monetary demands from clans who have recently received mining royalties and compensation payments. During Palia's lakisia I ran into Jepson, a teenager whom I had known since he was just a boy and who, like many young men his age, relished the excitement of tribal warfare. "Mi les long wokim gaden" (Working in the garden is boring), he told me, and went on to describe in intricate detail a daring nighttime raid he had recently been on that resulted in the burning of a large trade store and the killing of the store owner's son.

A few weeks before I interviewed Kangi, I was walking around the charred remains of the Porgera Elementary School that had been burned down in December 2005. From out of the surrounding forest a man appeared; he had a pair of binoculars and an AR-15 assault rifle. He stopped me, but after learning that I wasn't an enemy combatant but a university professor researching in Porgera, we went to his house for tea. He recounted how his clan—whose land the school was on—had received a significant compensation payment from the mine for some land destroyed during mining operations. A neighboring clan demanded some of the money, and when his clan refused, a fight broke out. Eventually, to disgrace his clan, the opposing clan burned down the elementary school. Almost a year later nothing had been rebuilt as the mine and the Porgera Development Authority were waiting for hostilities to be settled before embarking on the construction of a new elementary school. To make matters worse, the PDA building was ransacked and burned in 2006 by angry Porgerans after a drunken policeman (funded and housed by the mine) shot and killed a high school student.

Gun-based warfare has provided new benefits for men who seek their fortunes in the life market. As gun ownership is illegal in PNG, guns command great prices on the black market. Few individuals, let alone clans, can afford to own one. Kangi's clan had purchased an M-16 on the black market for K18,000 in 2004. With bullets costing between K20 and K30 each, clans that own guns train one man to use it and then hire these men out to other clans. These assassins are called Rambos (Wiessner and Pupu 2012) and command great respect in Porgeran society. Not only have they taken on the Rambo name, but men reported to me that before going out to fight they say, "Vietnam, one-man war," which they attribute to one of the Rambo movies. (And now that many of the children of my Porgeran friends are on Facebook, young men post doctored photos of themselves as soldiers with combat gear and M-16s.) The idea that the life market is envisioned as an individual struggle—a one-man war—also speaks to the transformation in group conflict from

Figure 7.7. New graves adjacent to the Enga Highway.

a collective endeavor to the alienating effects of individuals struggling to make do in a monetized economy.

A pile of rocks in the highway at the center of Apone Camp marked where Palia had been gunned down in the night. Traffic along the road—the constant flow of tanker trucks bringing diesel to the mining operations, PMVs, and flatbed Dyna trucks carrying people and vegetables to the markets—wove around the symbolic cairn, which in a few weeks would disappear as other events and circumstances erased the memories of Palia's death. Palia's clan members were erecting a more permanent monument to his killing, though, in the form of a multitiered grave platform adjacent to the highway. They were building it next to another multitiered grave painted in vibrant colors, complete with a flag pole flying the Papua New Guinean national flag. This grave held the remains of the LLG councilor killed by Kangi (figure 7.7). There were a number of these graves in Porgera, and I found them both disturbing and fascinating as, prior to the 2000s, graves were nondescript and rarely visible from public venues. Now, however, with the recent surges in violence related to mining wealth, people killed in these resource conflicts are buried in these multitiered, vividly painted graves adjacent to main roads and public areas to serve as reminders of the tragic consequences that mining has brought to many Porgerans.

Figure 7.8. Widows in mourning whose husbands have been killed in the life market.

Another change that comes from the incessant violence is the resurgence in "traditional" funeral garments worn by women. In 1999 and 2000 I had heard stories about what "women used to do in the past," which was to smear ashes on their face, wear disheveled and dirty reed skirts, and hang multiple strands of Job's tears (*Coix lacryma-jobi*) necklaces around their necks. Prior to 2006 I had never seen a woman wearing a widow's mourning garb. That year, however, I saw numerous widows dressed entirely in black, with ash-whitened faces and strand after strand of Job's tears encircling their necks (figure 7.8). While non-clan members are restricted from talking to mourning widows, I didn't need to ask them the intentions behind this new-old custom. In their silence, their mourning clothes spoke loudly to the pain and suffering generated by husbands who chose to work in the life market.

After I left Porgera, the fighting raged on between the Piango and the Wapena. Epe's son, George, would email me from his job at the mine and update me on the conflict. In April 2009, just before Operation Ipili commenced, George wrote, "There were 11 people killed just within 4 days and this terrible fight is still going on. All were buried but except for the Reserve police [man] is still in the Paiam morgue. My family was displaced by this last fighting and are living in the forest. I now sleep and work at the mine. The military is soon deploying to Porgera and there will be 300 personals [personnel] within Porgera to bring back law and order in the valley. The death records stands at 61 now in this fight and there are a total of 5 different fights now in Porgera." As a postscript to these conflicts, in early 2014 Ben posted a photo of himself with a pig on Facebook. I asked what it was for. Ben replied, "So many deaths to compensate. Sons and grandsons will continue compensating deaths. I would be appreciative if you were here to buy us a pig each time we compensate a death. I bought the pig you saw in the photo with me for K2,500. Pigs are very expensive here brother." Another young man contacted me through Facebook, writing, "The place where I was little and you did research [Kolatika in 1999], you would no longer recognize it. It is all bush now, covered in forests, and filled with the graves of the dead." A few weeks after I received George's email in 2009, Operation Ipili commenced, and over the next few months, in an attempt to stop the conflicts and oust illegal miners,[3] police and military personnel burned down over 130 homes, killed livestock, and beat and raped several Porgerans (Amnesty International 2010). Images in the press and the Amnesty International report showed scenes of burning houses and people standing in the midst of charred ruins. Since I started this book with images of burning, it seems apt to come full circle and describe more images of burning before moving on to draw some conclusions about the Porgera case. I think that my Porgeran friends would find this appropriate, as the burning of a garden always initiates the growth and development of new crops. And hopefully this book will contribute to finding new ways forward in Papua New Guinea's development boom.

Conclusion

Development, Resilience, and the End of the Land

Today's "developers" are like the alchemists of old who vainly tried to transmute lead into gold, in the firm belief that they would then have the key to wealth. The alchemists disappeared once it was realized that true wealth came from elsewhere—from people and trade. When will we realize that well-being does not come from growth?

GILBERT RIST, *The History of Development*

The Dilemmas of Development

The main idea that I have tried to articulate in this book is that resource development is alchemy. It is more a series of ritual procedures often doomed to failure than it is a science of human improvement. As such, development is a frequently critiqued field within the social sciences (Crush 1995; Ferguson 1994; Peet and Watts 1993; Sachs 1992). In *Seeing Like a State*, James Scott (1998) has argued the reason large-scale development projects fail is that bureaucratic processes simplify complex local practices and ignore the contributions that local knowledge can bring to bear on improving human lives (see also Escobar 2008). Building on this idea, Tania Li (2007) argues that development practices are largely concerned with "rendering technical" the issues they are attempting to address; that is, they separate expert from nonexpert knowledge and depoliticize development interventions (Ferguson 1994). In his book

Expectations of Modernity, James Ferguson (1999) describes the decline of Zambian mining development through the ways that mining frontiers in developing countries are marginalized by the global political economic system (see also Bridge 2001). He also details the sense of failed modernity that Zambians in the Copperbelt are experiencing due to the demise of the Zambian copper industry. Ferguson's work thus highlights the problems inherent in large-scale development planning that is dependent on the extraction of a global commodity subject to the whims of market mechanisms.

In a recent op-ed piece in the *New York Times*, the economist and Nobel laureate Paul Krugman explored three examples of what he called "money pits" or tales of "monetary regress—of the strange determination of many people to turn the clock back on centuries of progress." His first example is the Porgera Gold Mine, with its "terrible reputation for both human rights abuses . . . and environmental damage." Krugman writes that even such diametrically opposed economists as Adam Smith and John Maynard Keynes agreed that gold was merely a "dead stock" (Smith's words) tying up a nation's wealth. The mining of gold for wealth, he argues, is pointless, and moreover we have enough of it in vaults for our utilitarian purposes. Yet gold prices are now three times what they were a decade ago, Krugman concludes, "so dig they must."[1]

I have stressed that gold mining in Porgera is not driven solely by external forces seeking profit. There are of course international financiers and developers conjuring visions of wealth at play (Tsing 2000), but at the national level, mining contributes significantly to financing policies to pay off international loans and provide government services. At the provincial and local government levels, mining proceeds fund infrastructure projects and health and education initiatives. Locals themselves are desirous of the expected transformations that mining wealth will bring to their lives and communities. Nevertheless there are no examples of large-scale mining projects in PNG that satisfy all of the various parties involved, especially the local people who are hosts to the projects (Bainton 2008; Filer 1990; Golub 2014; Kirsch 2006). These projects typically fail on two fronts: when defining the local group is an exercise that turns dynamic and ever-changing social groups into static, fixed entities and when local communities misunderstand what kinds of transformations development will entail in their lives. Two contemporary works explore the tensions between development expectations and the environment in PNG. Working in the communities downriver from the world-class Ok Tedi copper and gold mine, Stuart Kirsch (2006) has argued that the Yonggom perceive development as "failed exchange." Analogous to Ferguson's conclusion, Kirsch (2006: 95) found that failed exchange is a form of "unrequited reciprocity" that centers the analysis not on "the performative contexts of exchange" but rather on those "who [have] been excluded or forgotten" in the exchange transaction. This exclusion from development has created multitudinous problems for both the Ok Tedi developers and the Yonggom from international attention and court settlements over environmental damages from mining. Paige West's (2006a) long-term research among the Gimi at the Crater Mountain Wildlife Management Area highlights the contentious and divergent expectations over development between conservation NGO workers and local communities. These tensions eventuated in the demise of conservation efforts around Crater Mountain as the Gimi aligned with mining companies who promised to bring them the development they hoped for but that conservation had not delivered in the ways they anticipated. This happened even though a number of the Gimi visited Porgera on a study tour (West 2006b). West relates the story of one of the men who went on the tour, who noted, "Look, the landowners, even the ones who got the good houses from the company, their forest is gone. Their mountain is gone. Their game is gone. Their clean water is gone" (306). While this was enough to initially convince the Gimi of the failures of mining development, they eventually changed their minds in the expectation that mining development would entail the building of hospitals, schools, and stores.

In essence I argue that Porgera is a massive development failure both socially and environmentally. While proceeds from mining development translate into forms of material improvement in the area, the costs of mining in human lives and the degradation of biodiversity far outweigh the benefits of development. This is the tension of alchemy: the notion that turning land into gold comes without great costs. Within the community the understanding of development is underpinned by discourses and expectations of progress and modernity. Porgeran conceptions of progress (variously construed in Tok Pisin as *develapmen*, "development"; *gutpela sindaun*, "quality of life"; *kamapim laip*, "improving life"; and *kisim sevises*, "receive services") articulate with the benefits that mining and a host of other quality-of-life issues associated with modernization are expected to bring. (For other PNG cases, see Halvaksz 2008; Knauft 2002; Kulick 1992; J. Robbins 2004; Robbins and Wardlow 2005; Stewart and Strathern 2002.) While Porgera has much more infrastructure development than many parts of highlands Papua New Guinea, enough Porgerans have traveled elsewhere to know that the region is still quite impoverished despite the presence of a world-class mine. In government circles, development operates within the same parameters of expectation that it does in the community, but it figures significantly in various individuals' attempts to control and gain prestige through the implementation of development programs. Whereas formerly men (and it was and still is mostly men) gained prestige by controlling the flow of pigs and shells in traditional exchange systems (Meggitt 1974; Strathern 1971) and by serving jail time and becoming acculturated to Australian colonialism, today they gain prestige and power by allocating money for road construction projects and the building of schools and aid posts.

The massive population increases into Porgera have exacerbated the social tensions and environmental problems associated with development such that violent conflicts are the norm. While the state has attempted in various ways to mediate these problems, it is challenged by political infighting among its various agents and institutions at local, provincial, and national levels. With the state unable to provide many of the government services Porgerans expect, Porgera Joint Venture often finds itself (unhappily, according to mining officials) fulfilling a quasi-state role through various community development initiatives. Thus PJV is split between engaging in technical, extractive mining practices and social, affiliative community development practices. Moreover increasing attention to the social and environmental impacts of mining on a global basis in the 1990s has forced PJV to pursue a "triple bottom line" corporate strategy,[2] where the focus is not just on economic profit but also on environmental sustainability and social development. Conversations with employees in the community relations branch of PJV highlighted that the kinds of social tensions that have developed in Porgera were never anticipated.[3] The mine's misunderstanding of social dynamics has further weakened its standing in the valley.

While development entails some *qualitative* transformation by the community and the government (such as an improved quality of life), this is a concern that is largely not shared by mining companies, despite shiny annual "sustainability" reports and corporate social responsibility publications saying otherwise. Mining development from the company's perspective is a *quantitative* endeavor; it focuses on tons of ore processed per day, ounces of gold produced per quarter, estimated and provable reserves of mineral ore, share prices, operating costs, and so forth. Reconciling the qualitative and quantitative dimensions of mining is what, for many Porgerans and outside observers, prompts their opinion of the

Porgera development process as a failure. However, from the perspective of shareholders and national and international financial institutions, Porgera is a resounding success. Since opening in 1990, the mine has produced over 18 million ounces of gold, which has contributed to about 9 percent of PNG's GDP and 13.4 percent of its export earnings on average over this time period.[4] Barrick Gold Corporation, the mine's operator, estimated at the end of 2013 that there were approximately 3.1 million ounces of proven and probable gold reserves remaining at Porgera.[5] At current gold prices, this is over $4 billion worth of gold still in the ground. Mining proceeds have been used for infrastructure improvements, such as schools, police posts, hospitals, and playing fields, in the valley and the province. Yet many Porgerans remain suspicious of the state's and the developer's intentions. These misunderstandings are more than just a standard formula of evil mining corporations, corrupt government officials, and naïve natives. The contentious politics of resource development are far more complex and situated within larger arguments about state-making and the rapid economic development of PNG, the environmental consequences of resource extraction, and the impacts of global capitalism and developmentalism (Sachs 1992; Ballard and Banks 2003).

With the advent of more mining in the Porgera region (Mt. Kare is expected to be a major mine in the near future) and more mining and petroleum development in the country, Papua New Guineans are going to have to find other mechanisms than the customary "landowners get all of the benefits" approach that has resulted in the social crises in Porgera. Moreover, with the environmental catastrophes that the Porgera, Panguna, and Ok Tedi mines have wrought on PNG's land and water resources, developers and politicians will need to explore ways of mitigating ecological ruin if mining is the future of economic development in PNG. The current path is neither socially nor ecologically sustainable (Filer and MacIntyre 2006).

Resilience, Ontology, and Social-Ecological Systems

Fikret Berkes and Carl Folke (2002: 125) argue that local resource management practices based on long-term yet dynamic ecological knowledge "put the brakes" on transformative changes (*sensu* Dwyer and Minnegal 2010) that increase vulnerabilities in social-ecological systems. Social memories and responses related to volcanic eruptions and major droughts and frosts support these claims. Porgerans and groups around them had incredibly complex mechanisms of exchange and social organization to ensure that people had access to a variety of different land types and altitudinal zones to foster livelihood resilience. The localization of mining benefits in the Special Mining Lease area has challenged this resilience, however, by concentrating people in the SML who hope to share in the wealth, while at the same time social networks have been truncated as SML landowners are loath to broadly share cash payments derived from mining. Before the mining era, endemic small-scale conflicts worked to put the brakes on transformation by continually reorganizing wealth in the society. With the advent of huge sums of money and the adoption of gun-based warfare, Porgeran conflicts have assumed a new kind of resilience in terms of being unable to halt the violence. Typically death compensations are expected to be paid within a few years, but with sons and grandsons having to pay the lingering compensation debts, anger over delayed compensation will certainly intensify the cycle of revenge killings spiraling through temporal and spatial frameworks in Porgeran society. Yet elsewhere in Enga Province people have been able to find collective measures to broker peace in warring communities (Wiessner and Pupu 2012).

I focused on Porgeran ontologies and their ecological philosophies to try to understand the ways that myths, attributions of intentionality to nonhumans, and the environment as sentient can shape understandings of novel events and provide resilient responses to dramatic changes. Holling and colleagues (2002b) examine myths and worldviews as part of a temporal scale along which resilience and change operate (see also Berkes 2008). The critical idea they offer is that adaptive cycles (as discussed in the introduction to part III), both spatial and temporal, are nested within one another. They refer to this as a "panarchy" as opposed to a hierarchy and use the example of a boreal forest to model these different scales. Increasing in spatial scale, one goes from a needle to a crown to a patch to a stand to a forest to a landscape. Spatial scale and temporal scale are linked such that smaller elements in the system are impacted by faster processes, while slower processes impact larger elements in the system to a greater extent. They explain:

> Small and fast scales are dominated by biophysical processes that control plant physiology and morphology. At the larger and slower scale of patch dynamics, interspecific plant competition for nutrients, light, and water influences local species composition and regeneration. At a still larger scale in a forest, meso-scale processes of

fire, storm, insect outbreak, and large-mammal herbivory determine structure and successional dynamics from tens of meters to kilometers, and from years to decades. At the largest landscape scales, climate, geomorphological, and biogeographical processes alter ecological structure and dynamics across hundreds of kilometers and over millennia. These processes produce patterns and are in turn reinforced by those patterns; that is, they are self-organized. (Holling et al. 2002b: 69)

Applying the concept of a panarchy to a social system, Holling et al. (2002b: 72) explore the relationships between "slowly developed myths," "faster rules and norms," and "still faster processes to allocate resources." While "slower levels constrain the behavior of faster levels," it is critical to recognize that "the attributes of the slower levels emerge from the experience of the faster" (72). Change in the system is generated "because productive novelty cascades up the levels, or because destructive catastrophes cascade down" (102). Locally derived ontologies and myths thus serve to guide resource management practices and access to resources.

Porgeran ontological systems serve to inhibit social transformation by providing templates with which to understand novel events. In essence they channel disruptive events into recognizable frameworks of experience (cf. Sahlins 1981). The experiences that Porgerans had with the Hagen-Sepik Patrol resulted in Taylor and Black being transformed into sky people and in the instantiation of a series of ritual efforts to ascend to the sky world and the promise of wealth and immortality that it promised. Gold and its origins were incorporated into myths detailing human reciprocal relations with python spirits that revitalized the environment and society. Local ontologies, myths, and the responses they engender serve an important role across Melanesia, and a number of scholars have explored the role they play in resilience—without actually using that terminology, though. Jeffrey Clark (1995: 380) points to Roy Wagner's (1975: 32) identification of so-called Melanesian cargo cults as a form of "reverse" anthropology. The cults are "Melanesian attempts to understand and appropriate an encompassing and dominant colonial power. In this scenario Europeans come to represent a known category . . . [and] are then subjected to a Melanesian epistemological framework and ritual engagement" (Clark 1995: 380). Clark argues that "'reverse' histories do exist—that is, non-Western forms of making history that interpret events over time not through a linear logic of cause and effect but through the interplay of myth, place, and consciousness" (381). Ontologies and myths in highlands societies work like an "epistemological lens" (397), in that they serve to focus and refract novel happenings, such as the coming of whites or the discovery of gold and oil, which can then be reconfigured in the structure of their own logics.

While many of the rituals guided by local ontologies have apparently declined, myths still serve as important sources for symbolic introspection and practical action. Dan Jorgensen (2014) has noted this as well for the Telefomin located some 200 kilometers west of the Porgera valley. The Telefomin, like the Porgerans, are embroiled in mining politics of their own due to their proximity to the Ok Tedi mine (Jorgensen 1990, 1996, 1998). Myths are important because there is a concern in these cultures with what Goldman (1983: 114) refers to as "first causes" for the Huli. In Huli, like Ipili and Enga, first causes are called *tene* (*tenge* in Enga), which can mean "source," "origin," "base," or "reason." Myths are critical in terms of this search for ultimate causality, even though it may not be linear, as Clark notes above. In support of this, Jorgensen (2001: 122) writes, "Currently known myths may contain details which offer clues to deeper meanings—*magam* [tene in Huli and Ipili]—that are significant for the world of experience. From this point of view, Telefol myth supports two kinds of innovation which can be deployed in the light of historical events, both of which are seen as part of pre-existing tradition. On the one hand, currently available myths may be given a new significance that was latent in features of the narratives themselves; on the other hand, new myths may emerge on condition that they articulate with previously known myths in some way." As Pamela Stewart and Andrew Strathern (2002: 183) argue, mine sites are central to this process of reinterpreting myth because of the "new transactions and productive processes that throw diverse sets of people together in relations of cooperation, struggles for advantage, and inequality" (see also Kirsch 2006; West 2006b). In their view myths both reflect the political ecology of these processes and provide a means for instigating action (Rumsey and Weiner 2001, 2004).

Local ontologies also help to inhibit ecological transformations. Until just forty years ago, Porgerans were constrained to living and gardening in a narrow altitudinal belt (andakama) between about 1,600 and 2,200 meters above sea level. Stories about spirits living in the realms above and below the andakama prevented humans from moving into these areas. Embedded in these stories were proscriptions about resource management and limitations on harvesting; the failure to properly manage resources would result in punishment by the guardian spirits of the rain forest. While fears of the lower rain forest spirit—the yu angini wanda—still structure Porgeran interactions with the lower rain forest, the same concerns have not been applied to the upper rain forest spirits—tawe wandakali (sky people). Part of this changed attitude is because each category of spirit has been incorporated into Christian forms. The yu angini wanda are perceived as demons controlled by Satan, whereas the tawe wandakali are thought of as angels. While tawe wandakali were always relatively benign, their resacralization has in some ways displaced them to living permanently in the sky world, spending less time in the upper rain forest than they would have traditionally. This displacement has allowed the movement of people into the upper rain forest, although there were other drivers, as discussed in chapters 6 and 7.

I emphasize these concerns with myth, place, and consciousness to demonstrate the contextuality and complexity of Porgeran social-ecological systems (cf. Morphy 1990, 1995). I also do so to suggest that it is the combination of resource development with particular ontological systems that ultimately determines the limits to resilience in these same systems. Resilience and vulnerability will always be shaped by the specific contexts of resource development, ontology, and ecology. This is what makes environmental anthropology, with its ability to work between scales of analysis yet remain grounded in an exposition of local lifeworlds, so well suited to studying resilience and vulnerability in social-ecological systems.

Gold and the End of the Land

Gold has a strange power over humanity. The Roman poet Ovid (2010: 298) wrote of the Greek (and Roman) myth of King Midas, who was granted a wish by Dionysius (Bacchus) that whatever he touched would turn to gold:

> The god nodded and gave him the harm he asked for,
> Sorry he hadn't seen fit to request something better.
> Our hero went off pleased with his fatal gift
> And tried by touching things one at a time.
> Not really believing, he broke off a green twig
> From a hanging oak branch. The twig became gold.
> He picked up a stone, and it blanched into gold.
> He touched a clod of earth, and at his potent touch
> It became a lump of gold. . . . His mind

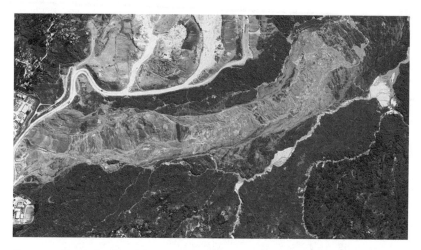

Figure C.1. The confluence of the Porgera River with mining tailings. Source: Google Earth.

could hardly contain his hopes, imagining
That everything was gold.

King Midas's delight soon turned to horror as he realized that even the food and drink he tried to put into his mouth were transformed into gold. Realizing that his golden gift would end in his starvation, Midas pleaded with Dionysius to take away his power. Dionysius heard his entreaties and told Midas to go wash in the River Pactolus and that his power would be released by the water. Midas did so, turning the river sands to gold.

We would be wise to heed the warning in the myth of King Midas, whose touch turned nature to gold. Our quest for the wealth gold can bestow drives capital to the ends of the earth for this soft, lustrous metal. In its wake it often leaves ruined landscapes and fractured communities. The Porgera River has not been as lucky as the Pactolus. Iron oxides in the tailings waste have turned the Porgera blood red. On any given day in Porgera, a bloody braid of watery tailings snakes across the gray glacier of waste rock to mix into the main body of the Porgera River (figure C.1). Even several kilometers downstream of the tailings disposal site the river remains red. Porgerans perceive the *redpela wara* ("red water" in Tok Pisin) with the same trepidation that they do the polluting power of menstrual blood (Biersack 2006a). Just as the Huli equated gold from Mt. Kare with feminine pollution due to the breakdown of gender segregation norms during the Mt. Kare gold rush (Clark 1993), Porgerans interpret the pollution of gold in both a metaphorical, feminine sense and an actual, material sense.

The pollution of the Porgera River aptly highlights the metabolic rift through which capitalist resource extraction works to sunder humans from their environment. That we can sacrifice a river and a landscape and not detrimentally impact the people whose lives are tied to them is a fallacy of resource extraction in Porgera and other places the world over. Parallel to the metabolic rift is what Aldo Leopold (1949: 202) called the lack of a land ethic in capitalist societies: "An ethic, ecologically, is a limitation on freedom of action in the struggle for existence. An ethic, philosophically, is a differentiation of social from anti-social conduct. These are two definitions of one thing." Whereas ethics have been developed for relationships between individuals, and between individuals and society, Leopold argues that the same ethic has not been developed for the land: "The land relation is still strictly economic, entailing privileges but not obligations" (203). The call for a land ethic and the recognition of a metabolic rift in capitalism highlight the notion that land (or nature) has no agency, that we don't have to engage with it in the same terms as we do individuals or society. As I have highlighted throughout this book, however, the agency of the land is a foundational concept in Porgeran worldview (cf. Halvaksz 2003). It consists of skin (yu umbuaini) and bone (yu kulini) and harbors a life force (yu ipane) essential to people, spirits, flora, and fauna. Even one of the products of this land—gold—is an "uncooperative" resource (cf. Bakker 2004). From the early prospectors who complained of its location in boulder-choked gorges, to Porgeran myths that depict a landscape riddled with gold yet unattainable due to the secretive movements of python spirits, gold defies human efforts to capture it completely (which is even reflected in quarterly mining reports documenting the percentage recoverable from each mine). Gold also has an agency in its ability to structure social differentiation and economic inequality—why it appears in one place and not another. Its ability to transform human social-ecological systems in unexpected ways, in Porgera and elsewhere, is the alchemy of global resource extraction. While I have used the notion of alchemy as an interpretive device to explore transformations in Porgeran lands and lives, it also serves to capture the very contradictions of modernity that development is supposed to bring to its intended audience. As such it represents the limits of the scientific enterprise through which people's lives can be improved.

At the beginning of chapter 1, I asked what it meant for a horticultural society to call a gold mine a garden—to "eat" gold. Eating in Porgera ensures the flow of ipane (grease) through the social-ecological system. Eating delineates the reciprocal relations between the people, the spirits, and the land. But gold has no "grease" in the contemporary era; it comes from wealth without work and is an antiproduct that, as in King Midas's case, is inedible in a society where conviviality shapes social forms (see also Halvaksz 2006). And yet Porgerans use the same term, *eating,* to describe what they do with the money that comes from mining development. However, this form of eating is similar to the way that spirits "eat" people. It is antisocial—individualistic and short term—and not put into the kinds of social projects that link people and groups across time and space. What gold mining has done and what Marx (1990: 638) argues industrial agriculture did to the land is rob the land of its productivity by "undermining the original sources of all wealth—the soil and the worker."

Eating the wealth from gold is analogous to the way that mining eats into the skin and bone of the Porgeran landscape. This is why Porgerans reflect on the changed circumstances of their lives as yu koyo peya (the end of the land). The end of the land is the end of the reciprocal ties between the people, the land, and the spirits. Moreover, with the advent of Christianity, the land is no longer the source of ritual focus. The end of the land—whether interpreted through tradition, where the yu kimbuni (earth joint) spirits leave their sacred spots and meet in the high mountains behind Porgera and dance and drum the world to pieces, or interpreted through Christian end times scenarios—promises salvation from a world no longer amenable to human manipulation. The destructive forces of globalization, such as the social and environmental impacts of mining, Christian apocalypse, and a burgeoning firearms market, found fertile ground in local ontologies in Porgera, with entropy and creative destruction as central philosophical concepts. But there is a glimmer of resilience in people's attitudes, as Kipan Wuambo said to me in the last days of December 1999, at the height of apocalyptic fervor in Porgera: "Will the ground end in 2000? Will it end in 2001? I don't know, but I do know that it will end sometime. But until then, I'll make gardens and eat sweet potato, I'll make exchanges and eat pig. Maybe a mine will open in Tipinini and I'll eat money. But then who knows, maybe we'll all just go back to living like our ancestors did—no mine, no cars, no rice, no money." Then Kipan laughed heartily at the contradictions and transitory nature of life and wealth in a resource frontier.

Introduction

1. In 2000 the mine was planning on beginning shutdown procedures in 2006 due to the low price of gold.

Chapter 1. Resource Frontiers in the Montane Tropics

1. Harvard Law School also investigated this incident. As a result of their findings and Amnesty International's, the Norwegian government divested all of its pension holdings of Barrick Gold Inc. stock.

2. The United Nations Collaborative Programme on Reducing Emissions from Deforestation and Forest Degradation in Developing Countries was introduced at the eleventh Conference of the Parties to the UN Framework Convention on Climate Change by the governments of Papua New Guinea and Costa Rica in 2005 and adopted at the thirteenth conference in Bali in 2007. Two of the major international carbon initiatives are the World Bank's Global Forest Carbon Partnership and the Australian government's International Forest Carbon Initiative.

3. Rowan Callick, "Miners Hope to Restart Bougainville Gold and Copper Mine," *Australian*, December 28, 2010.

4. Rainfall data were provided by the Environmental Division of Placer Pacific, Ltd., the operating partner for the Porgera Joint Venture gold mine as of 2004, when I received the data.

5. My use of the term *primary rain forest* does not mean that I believe these are "virgin" forests (Willis et al. 2004) that are untouched by people. Rather these

are most likely very old fallow forests, especially in the midaltitude zone. Moreover, as shown later in the chapter, the low biodiversity of primary mid-altitude forests indicates that they are most likely recovering from or still experiencing some amount of anthropogenic disturbance. *Primary rain forest* is the translation I have chosen for the Ipili word *ima* (forest reserve) as it is used in the context of comparing this land cover type to *e pale* (old garden, i.e., second-growth forest).

6. Pronounced *ip-a-lee*, accent on the first syllable with the first i pronounced like the *i* in *pin*, the second *i* like the *a* in *sofa*, and the third *i* like the vowel sound in the name *Lee*.

7. Countrywide the annual growth rate is 2.2 percent according to the United Nations.

Chapter 2. Colonialism, Mining, and Missionization

1. Papers of John Black, National Library of Australia, Canberra, MS 8346/3/9.
2. Papers of Jim Taylor, National Library of Australia, Canberra, MS 9218/5/4 (hereafter cited parenthetically in the text).
3. Papers of John Black, National Library of Australia, Canberra, MS 8346/3/9. All of the other quotes in this chapter are from his second diary, MS 8346/3/11.
4. Wabag patrol report #2 1944/45, National Archives of Papua New Guinea, Port Moresby. Patrol reports (PR) from the various patrol posts are hereafter cited parenthetically in the text.
5. "Gold Seekers Disappointed: New NG Field Now Called a Washout," *Pacific Islands Monthly*, May 1948, 9.
6. Mining prospects were also increased by the demise of the gold standard in the early 1970s, which allowed the price of gold to rise from its fixed exchange rate of $35 per ounce to over $800 per ounce by 1980.

Part II. Indigenous Philosophies of Nature, Culture, and Place

1. What in ecological terms would be referred to as the difference between their "fundamental" and "realized" niches (Hutchinson 1965; see also Hardesty 1975).

Chapter 3. Land. *Yu*

1. Technically land "ownership" occurs at the subclan level, but Porgerans often talk about it as if it were occurring at the clan level.
2. The exchange rate in 1989 was US$1.17 to K1.
3. The souls of men and women travel down different roads on the way to the mountain Kewalieme in Tamakale. Men travel down the *akali palane asini*, which is a road lined with a kind of cane grass called *yangetene*. Women travel down the *wanda palane asini*, which is lined with *uta* sedge grass (*Eleoecharis dulcis*) used for making grass skirts.

Chapter 4. People: Wandakali

1. Interestingly Sillitoe (2010: 354) reports that the nearby Wola have no word for "work."
2. In Porgera *ita* is the word for "tree," whereas in Paiela they use *sia*.
3. Fallow lands in Porgera, unlike many other places in the highlands, tend to become overgrown with trees and not cane grass (*Miscanthus floridulus*). Some areas do become covered with cane grass, the removal of which is very labor-intensive.
4. A pseudonym. In some cases, especially concerning violence and killings, I have changed people's names. I have tried to capture some of the sense of their name in the pseudonym, and if their name was in Ipili or Enga, I have chosen another Ipili or Enga name. Biblical names have been substituted for other biblical names. In the case of September and some others, I have tried to represent the fact that they had a name in English that isn't commonly (if ever) used by Euro-Americans.
5. Peter Muyu explained it as follows: "Yu mas lukim ol lain sapos yu laik kisim kompenseshun, graun pe, o wanem samting" [You need to go and be social with the whole clan if you expect to receive compensation, land occupation fees, or something; field notes, December 1998].

Chapter 5. Spirits: Yama

1. A similar word is *itane*, used for the trunk of a banana or pandanus tree. *Ita* is "tree"; hence bananas and pandanus are "things that are like a tree."
2. Some storytellers give Lemeane's name as Popaya-Akali-Lemeane, which translates as "cloud-man-Lemeane," and refer to Lakeame by the name Tambu-Wana-Ipali, which translates as "tree fern-woman-Ipali." Clouds and tree ferns (of the upper rain forest) are associated with sky people.
3. The sisters' names are also derived from the words *lipi* (nearby at a higher elevation) and *napi* (nearby at a lower elevation; see chapter 3). In one of the many versions of this myth that Biersack (1999b) heard, after killing Taiyundika one of the sisters flees to the Porgera valley and marries Kupiane.
4. In 2012, three districts in Southern Highlands Province became the new Hela Province.
5. In a previous article (Jacka 2001), I wrote that the clan pools were called *ipa angini* (body of water). Technically this is correct, as all lakes and pools are called ipa angini, but the more definitive term for a pool that contains the spirits of deceased ancestors is *ipa ne*.

Part III. Social-Ecological Perturbations and Human Responses

1. Paul Basken, "That Elastic Term," *Chronicle of Higher Education*, May 6, 2013.

Chapter 6. Ecological Perturbations and Human Responses

1. Blong (1982: 90–91) reports on one tephra layer above the Tibito Tephra in the upper Wage area, near the sacred site of Tuandaka, indicating a later volcanic event than the Long Island eruption. However, the tephra layer is very localized and found nowhere outside of the immediate upper Wage basin. Neither the source nor the age of the tephra layer could be determined, although it was "clearly younger than Tibito Tephra," and Blong labeled it "an enigma." This event could reflect the source of Undupi's name, which was given to him at birth sometime around the mid-1800s.

2. The Southern Oscillation Index (SOI) is a standardized measurement of the surface sea level pressure difference between Tahiti and Darwin, Australia. Negative SOI numbers are generally associated with El Niño conditions. The other aspect of ENSO is the Oceanic Niño Index, which is based on a threshold of +/- 0.5°C changes in sea surface temperatures in the Niño 3.4 region (5°N–5°S, 120°E–170°W); to qualify as an El Niño or La Niña, the threshold must be met for a minimum of five consecutive overlapping seasons.

3. After the killings, Ipayema, the cult leader, was killed. Timbapu escaped and went to his mother's people in Wage. His wife and children stayed in Porgera. In the 1980s Timbapu returned to Porgera and was killed in a tribal fight "at almost the same spot where Ipayema had been killed," according to his son, Korokan.

4. Porgerans themselves do not cite the 1941 date but mention a drought that occurred prior to the cults that took place in the mid-1940s. The 1941 date comes from wartime observations by Australian forces in PNG (see chapter 2).

5. National Oceanic and Atmospheric Administration, Climate Prediction Center, http://www.cpc.ncep.noaa.gov/products/analysis_monitoring/ensostuff/ensoyears.shtml. To create the index, I standardized seasonal rainfall with Z scores; the observed rainfall from the three-month mean was subtracted from the mean for that period and the result was divided by the standard deviation of that period (Ali and Lebel 2008). For example, January, February, and March (JFM; February, March, and April [FMA]; March, April, and May [MAM], etc.) rainfall amounts were averaged to provide an overall mean and standard deviation for twelve three-month periods each year over the years 1974 to 2011.

6. I defined normal climate months as periods when the three-month running mean rainfall averages were between 0.5 and -0.5 standard deviations from the overall mean for those three months. Normal climate months were scored 0. I scored wet months as follows: 0.5 for standard deviations between 0.5 and 0.99, 1.0 for standard deviations between 1.0 and 1.49, and so forth. I scored dry months the same way, although they received a negative score. To classify as an extreme wet or dry period, four consecutive months had

to have an absolute score greater than 0.5 for each month, with at least two months greater than one standard deviation. I then summed the modified index scores for each month to indicate the intensity of the wet or dry period.

7. Very little is known of income levels in Porgera. In a survey of 122 households in 1992, Glenn Banks (1999: 115) found the average monthly income to be K367.06. However, most of the households he surveyed were among landowning communities. In a 1999 survey I conducted of forty households in Tipinini, the average household monthly income was K62.

8. In Tipinini the 1957 census listed 644 people, the 1959 census listed 622 people, and a May 1960 census listed 721 people. Given equivalent variance of about 14 percent for the other census areas, it is assumed that populations in the upper Porgera were still less than in the other areas. The variance in percentage of people living in the lower Porgera is due to uncertainty about where to put the 1957 population of Paiam. Oral history and patrol reports indicate there were no people living in the present census unit of Paiam prior to the 1960s. If the Paiam census of 1957 is included in the upper Porgera census, then only 62 percent of the population lived in the lower Porgera. If included in the lower Porgera census, then the percentage of people living in the lower Porgera increases to 74.9.

9. Totals in 1980 and 1990 do not match the official totals from the PNG census. The 1980 census lists 5,013 people in Porgera, while the 1990 census lists 9,255. The numbers in table 6.1 are from the raw census figures showing people per census unit.

10. Image acquisition dates were August 19, 1972, August 27, 1982, September 29, 1991, December 8, 2002, and July 7, 2013.

11. Due to temporal and budgetary constraints, voucher specimens were not collected for determination of botanical species in 2006. Species names were determined by previous work from other researchers and comparison of cognate terms among neighboring groups. In the 1970s a dictionary of the Ipili language was compiled by Dr. Frances Ingemann, and vernacular tree names were recorded and identified in the field by John Womersley, the assistant director in the Division of Botany, Department of Forests, PNG (Ingemann n.d). As well, Ipili and Enga share cognates for several tree species with neighboring ethnolinguistic groups (Huli and Wola), among whom much species identification work has already been done (Ballard 1995; Sillitoe 1996, respectively). From the entire sample, for all species with five or more individuals, there are only seven species whose scientific names were not determined.

12. Importance value = relative frequency + relative density + relative dominance. Relative frequency = frequency of one species/total frequency of all species * 100. Relative density = density of one species/total density of all species * 100. Relative dominance = basal area of one species/total basal area of all species * 100.

13. Simpson's diversity is lower in midelevation secondary plots than in high-elevation secondary plots as Simpson's tends to emphasize the more common species in a sample. Midelevation secondary forests contain numerous rare species that the other diversity measures highlight.

Chapter 7. Social Dislocations

1. My goal here is not to delve into some arcane argument about the labor theory of value from a Marxist perspective. Rather I merely want to highlight that when Porgerans evaluate wealth items, especially pigs, they inevitably mention all of the "hard work" that went into their production. This is true of all items that may not fit a strict definition of "wealth" in the sense that they have exchange value: abundant gardens, well-built houses, finely crafted net bags, and so forth.

2. Both had been elected in the 1997 elections.

3. Porgera is notorious in the press and corporate publications for the huge number of illegal miners, people who enter the open pit and try to smuggle gold out of the SML area. Barrick Gold officials estimate at times over eight hundred people per day enter the open pit. Papua New Guinean and Australian media outlets regularly feature stories of killings of illegal miners by security officials working for Barrick Gold. It is an intractable issue and one that will only worsen with continuing economic disparities in the region.

Conclusion

1. Paul Krugman, "Bits and Barbarism," *New York Times*, December 22, 2013.

2. In 1999, under increasing public pressure to operate with greater transparency, nine multinational mining firms formed the Global Mining Initiative in preparation for the 2002 World Summit on Sustainable Development (the tenth anniversary of the Rio Earth Summit). A two-year initiative (from 2000 to 2002) called Mining, Minerals, and Sustainable Development was commissioned to study the challenges of pursuing sustainable development in the mining industry. As a result of this work, which was undertaken by the International Institute for Environment and Development, nearly all of the major mining companies are members of the International Council on Mining and Minerals, which seeks "to advance their commitment to sustainable development" (http://www.icmm.com/about-us/our-history). At least in Porgera's case, whether any of these initiatives and organizations has reduced environmental damages and social disruptions is highly debatable.

3. Several PJV employees in community relations told me that their department went from being one of the smallest at the beginning of mining to the largest by 1999. I had no way of validating whether or not this was

true, but it was at least what the community relations people themselves believed.

4. Mining production records and the economic contribution of mining for Porgera and PNG were obtained from the Porgera Environmental Advisory Komiti (PEAK) website (http://www.peakpng.org) and from annual country reports from the International Monetary Fund's website. PEAK oversees environmental issues with the Porgera Mine (see Coumans 2011).

5. "Porgera (95 Percent)," Barrick, http://barrick.com/operations/papua-new -guinea/porgera/default.aspx.

REFERENCES

Adger, W. Neil, et al. 2009. "Are There Social Limits to Adaptation to Climate Change?" *Climatic Change* 93: 335–54.

Agrawal, Arun. 2003. "Sustainable Governance of Common-Pool Resources: Context, Methods, and Politics." *Annual Review of Anthropology* 32: 243–62.

Agrawal, Arun. 2005. *Environmentality: Technologies of Government and the Making of Subjects.* Durham, NC: Duke University Press.

Agrawal, Arun, and C. Gibson. 1999. "Enchantment and Disenchantment: The Role of Community in Natural Resource Conservation." *World Development* 27 (4): 629–49.

Ali', Abdou, and Thierry Lebel. 2008. "The Sahelian Standardized Rainfall Index Revisited." *Journal of Climatology* 29 (12): 1705–14.

Allen, B., R. Michael Bourke, and L. Hanson. 2001. "Dimensions of PNG Village Agriculture." In *Food Security of Papua New Guinea: Proceedings of the Papua New Guinea Food and Nutrition 2000 Conference*, ed. R. M. Bourke, M. G. Allen, and J. G. Salisbury, 529–53. Canberra: Australian Centre for International Agricultural Research.

Allen, Bryant. 1989. "Frost and Drought through Time and Space, Part I: The Climatological Record." *Mountain Research and Development* 9 (3): 252–78.

Allen, Bryant, and Stephen Frankel. 1991a. "Across the Tari Furoro." In *Like People You See in a Dream: First Contact in Six Papuan Societies*, ed. E. Schieffelin and R. Crittenden, 88–124. Stanford, CA: Stanford University Press.

Allen, Bryant, and Stephen Frankel. 1991b. "The Historical Impact: The Huli." In *Like People You See in a Dream: First Contact in Six Papuan Societies*, ed. E. Schieffelin and R. Crittenden, 268–73. Stanford, CA: Stanford University Press.

Amnesty International. 2010. *Undermining Rights: Forced Evictions and Police Bru-tality around the Porgera Gold Mine, Papua New Guinea*. London: Amnesty In-ternational.

Appadurai, Arjun. 1996. *Modernity at Large: Cultural Dimensions of Globalization.* Minneapolis: University of Minnesota Press.

Ash, Julian. 1982. "The *Nothofagus* Blume (Fagaceae) of New Guinea." In *Biogeogra-phy and Ecology of New Guinea*, ed. J. L. Gressitt, 355–80. The Hague: Dr. W. Junk.

Bainton, Nicholas A. 2008. "The Genesis and the Escalation of Desire and An-tipathy in the Lihir Islands, Papua New Guinea." *Journal of Pacific History* 43 (3): 289–312.

Bakker, Karen. 2004. *An Uncooperative Commodity: Privatizing Water in England and Wales.* Oxford: Oxford University Press.

Bakker, Karen, and Gavin Bridge. 2006. "Material Worlds? Resource Geographies and the 'Matter of Nature.'" *Progress in Human Geography* 30 (1): 5–27.

Balée, William. 2006. "The Research Program of Historical Ecology." *Annual Review of Anthropology* 35: 75–98.

Balée, William, and David G. Campbell. 1990. "Evidence for the Successional Status of Liana Forest (Xingu River Basin, Amazonian Brazil)." *Biotropica* 22 (1): 36–47.

Ballard, Chris. 1995. "The Death of a Great Land: Ritual, History and Subsistence Revolution in the Southern Highlands of Papua New Guinea." PhD diss., Aus-tralian National University.

Ballard, Chris. 1998. "The Sun by Night: Huli Moral Topography and Myths of a Time of Darkness." In *Fluid Ontologies: Myth, Ritual and Philosophy in the High-lands of Papua New Guinea*, ed. L. R. Goldman and C. Ballard, 67–86. Westport, CT: Bergin and Garvey.

Ballard, Chris. 2000. "The Fire Next Time: The Conversion of the Huli Apocalypse." *Ethnohistory* 47: 205–26.

Ballard, Chris, and Glenn Banks. 2003. "Resource Wars: The Anthropology of Mining." *Annual Review of Anthropology* 32: 287–313.

Banivanua-mar, Tracey. 2007. *Violence and Colonial Dialogue: Australia-Pacific Inden-tured Labor Trade.* Honolulu: University of Hawai'i Press.

Banks, Glenn. 1999. "The Economic Impact of the Mine." In *Dilemmas of Develop-ment: The Social and Economic Impact of the Porgera Gold Mine, 1989–1994*, ed. C. Filer, 88–127. Boroko, PNG: National Research Institute.

Banks, Glenn. 2003. "Landowner Equity in Papua New Guinea's Minerals Sector: Review and Policy Issues." *Natural Resources Forum* 27: 223–34.

Banks, Glenn, and Chris Ballard. 1997. *The Ok Tedi Settlement: Issues, Outcomes, and Implications.* Canberra: National Centre for Development Studies, Research School of Pacific Studies, Australian National University.

Barrick Gold Inc. 2010. *Porgera Joint Venture Information Booklet 2010.* Toronto: Bar-rick Gold Inc.

Basso, Keith H. 1996a. *Wisdom Sits in Places: Landscape and Language among the Western Apache.* Albuquerque: University of New Mexico Press.

Basso, Keith H. 1996b. "Wisdom Sits in Places: Notes on a Western Apache Landscape." In *Senses of Place*, ed. S. Feld and K. H. Basso, 53–90. Santa Fe, NM: School of American Research Press.

Bayliss-Smith, Tim. 1996. "People-Plant Interactions in the New Guinea Highlands: Agricultural Heartland or Horticultural Backwater." In *The Origins and Spread of Agriculture and Pastoralism in Eurasia*, ed. D. R. Harris, 499–523. London: University College London.

Bayliss-Smith, Tim, Edvard Hviding, and T. C. Whitmore. 2003. "Rainforest Composition and Histories of Human Disturbance in Solomon Islands." *Ambio* 32 (5): 346–52.

Bebbington, A. J., and Simon Batterbury. 2001. "Transnational Livelihoods and Landscapes: Political Ecologies of Globalization." *Ecumene* 8 (4): 369–80.

Bell, Joshua. 2006. "Marijuana, Guns, Crocodiles and Radios: Economies of Desire in the Purari Delta." *Oceania* 76 (3): 220–34.

Bell, Joshua. 2009. "Documenting Dissent: Struggles for Recognition in the Purari Delta of Papua New Guinea." *Australian Journal of Anthropology* 20 (1): 28–47.

Bender, Barbara, ed. 1993. *Landscape: Politics and Perspectives*. Oxford: Berg.

Berkes, Fikret. 2008. *Sacred Ecology*. New York: Routledge.

Berkes, Fikret, Johan Colding, and Carl Folke. 2000. "Rediscovery of Traditional Environmental Knowledge as Adaptive Management." *Ecological Applications* 10 (5): 1251–62.

Berkes, Fikret, Johan Colding, and Carl Folke. 2003a. Introduction to *Navigating Social-Ecological Systems: Building Resilience for Complexity and Change*, ed. F. Berkes, J. Colding, and C. Folke, 1–29. Cambridge: Cambridge University Press.

Berkes, Fikret, Johan Colding, and Carl Folke, eds. 2003b. *Navigating Social-Ecological Systems: Building Resilience for Complexity and Change*. Cambridge: Cambridge University Press.

Berkes, Fikret, and Carl Folke. 2000a. "Linking Social and Ecological Systems for Resilience and Sustainability." In *Linking Social and Ecological Systems: Management Practices and Social Mechanisms for Building Resilience*, ed. F. Berkes and C. Folke, 1–25. Cambridge: Cambridge University Press.

Berkes, Fikret, and Carl Folke, eds. 2000b. *Linking Social and Ecological Systems: Management Practices and Social Mechanisms for Building Resilience*. Cambridge: Cambridge University Press.

Berkes, Fikret, and Carl Folke. 2002. "Back to the Future: Ecosystem Dynamics and Local Knowledge." In *Panarchy: Understanding Transformations in Human and Natural Systems*, ed. L. Gunderson and C. S. Holling, 121–46. Washington, DC: Island Press.

Berlin, Brent, Dennis Breedlove, and Peter Raven. 1973. "General Principles of Classification and Nomenclature in Folk Biology." *American Anthropologist* 75 (1): 214–42.

Biersack, Aletta. 1987. "Moonlight: Negative Images of Transcendence in Paiela Pollution." *Oceania* 57: 178–94.

Biersack, Aletta. 1991. "Prisoners of Time: Millenarian Praxis in a Melanesian Valley." In *Clio in Oceania: Toward a Historical Anthropology*, ed. A. Biersack, 231–96. Washington, DC: Smithsonian Institution Press.

Biersack, Aletta. 1995a. "Heterosexual Meanings: Society, Economy, and Gender among Ipilis." In *Papuan Borderlands: Huli, Duna, and Ipili Perspectives on the Papua New Guinea Highlands*, ed. A. Biersack, 231–64. Ann Arbor: University of Michigan Press.

Biersack, Aletta. 1995b. "Introduction: The Huli, Duna, and Ipili Peoples Yesterday and Today." In *Papuan Borderlands: Huli, Duna, and Ipili Perspectives on the Papua New Guinea Highlands*, ed. A. Biersack, 1–56. Ann Arbor: University of Michigan Press.

Biersack, Aletta. 1996a. "'Making Kinship': Marriage, Warfare, and Networks among Paielas." In *Work in Progress: Essays in New Guinea Highlands Ethnography in Honour of Paula Brown Glick*, ed. H. LeVine and A. Ploeg, 19–42. Frankfurt: Peter Lang.

Biersack, Aletta. 1996b. "Word Made Flesh: Religion, the Economy and the Body in the Papua New Guinea Highlands." *History of Religions* 36: 85–111.

Biersack, Aletta. 1998a. "Horticulture and Hierarchy: The Youthful Beautification of the Body in the Paiela and Porgera Valleys." In *Adolescence in Pacific Island Societies*, ed. G. Herdt and S. Leavitt, 71–91. Pittsburgh: University of Pittsburgh Press.

Biersack, Aletta. 1998b. "Sacrifice and Regeneration among Ipilis: The View from Tipinini." In *Fluid Ontologies: Myth, Ritual and Philosophy in the Highlands of Papua New Guinea*, ed. L. R. Goldman and C. Ballard, 43–66. Westport, CT: Bergin and Garvey.

Biersack, Aletta. 1999a. "Introduction: From the 'New Ecology' to the New Ecologies." *American Anthropologist* 101: 5–18.

Biersack, Aletta. 1999b. "The Mount Kare Python and His Gold: Totemism and Ecology in the Papua New Guinea Highlands." *American Anthropologist* 101: 68–87.

Biersack, Aletta. 2001. "Reproducing Inequality: The Gender Politics of Male Cults in Melanesia and Amazonia." In *Gender in Amazonia and Melanesia: An Exploration of the Comparative Method*, ed. T. Gregor and D. Tuzin, 69–90. Berkeley: University of California Press.

Biersack, Aletta. 2004. "The Bachelors and Their Spirit Wife: Interpreting the Omatisia Ritual of Porgera and Paiela." In *Women as Unseen Characters: Male Ritual in Papua New Guinea*, ed. P. Bonnemere, 98–119. Philadelphia: University of Pennsylvania Press.

Biersack, Aletta. 2005. "On the Life and Times of the Ipili Imagination." In *The Making of Global and Local Modernities in Melanesia: Humiliation, Transformation and the Nature of Cultural Change*, ed. J. Robbins and H. Wardlow, 135–62. Burlington, VT: Ashgate.

Biersack, Aletta. 2006a. "Red River, Green War: The Politics of Nature along the Porgera River." In *Reimagining Political Ecology*, ed. A. Biersack and J. B. Greenberg, 233–80. Durham, NC: Duke University Press.

Biersack, Aletta. 2006b. "Reimagining Political Ecology: Culture/Power/History/ Nature." In *Reimagining Political Ecology*, ed. A. Biersack and J. B. Greenberg, 3–40. Durham, NC: Duke University Press.

Biersack, Aletta. 2011a. "The Sun and the Shakers, Again: Enga, Ipili, and Somaip Perspectives on the Cult of Ain, Part One." *Oceania* 81 (2): 113–36.

Biersack, Aletta. 2011b. "The Sun and the Shakers, Again: Enga, Ipili, and Somaip Perspectives on the Cult of Ain, Part Two." *Oceania* 81 (3): 225–43.

Bird-David, Nurit. 1990. "The Giving Environment: Another Perspective on the Economic System of Gatherer-Hunters." *Current Anthropology* 31 (2): 189–96.

Blaikie, Piers, and Harold Brookfield. 1987a. "Defining and Debating the Problem." In *Land Degradation and Society*, ed. P. Blaikie and H. Brookfield, 1–26. New York: Methuen.

Blaikie, Piers, and Harold Brookfield, eds. 1987b. *Land Degradation and Society.* New York: Methuen.

Blong, Russell J. 1982. *The Time of Darkness: Local Legends and Volcanic Reality in Papua New Guinea.* Seattle: University of Washington Press.

Bonnell, Susanne. 1999. "Social Change in the Porgera Valley." In *Dilemmas of Development: The Social and Economic Impact of the Porgera Gold Mine, 1989–1994*, ed. C. Filer, 19–87. Boroko, PNG: National Research Institute.

Bougainville Copper Limited. 2013. Annual Report, 2012. Accessed February 27, 2014. http://www.bcl.com.pg/wp-content/uploads/2013/03/Annual-Report-Release-March-13.pdf.

Bourdieu, Pierre. 1977. *Outline of a Theory of Practice.* Cambridge: Cambridge University Press.

Bourke, R. Michael. 1989. "The Influence of Soil Moisture on Sweet Potato Yield in the Papua New Guinea Highlands." *Mountain Research and Development* 9 (3): 322–28.

Bourke, R. Michael, and Tracy Harwood, eds. 2009. *Food and Agriculture in Papua New Guinea.* Canberra: ANU E Press, Australian National University.

Braun, Bruce. 2002. *The Intemperate Rainforest: Nature, Culture, and Power on Canada's West Coast.* Minneapolis: University of Minnesota Press.

Bridge, Gavin. 2001. "Resource Triumphalism: Postindustrial Narratives of Primary Commodity Production." *Environment and Planning* A 33: 2149–73.

Brookfield, Harold, and Paula Brown. 1963. *Struggle for Land: Agriculture and Group Territories among the Chimbu of the New Guinea Highlands.* Melbourne: Oxford University Press.

Brosius, J. Peter. 1997. "Endangered People, Endangered Forest: Environmentalist Representations of Indigenous Knowledge." *Human Ecology* 25: 47–69.

Brown, M., and Jocelyn M. Powell. 1974. "Frost and Drought in the Highlands of Papua New Guinea." *Journal of Tropical Geography* 38: 1–6.

Bryant, Raymond. 1992. "Political Ecology: An Emerging Research Agenda in Third-World Studies." *Political Geography* 11 (1): 12–36.

Bulmer, Ralph. 1967. "Why Is the Cassowary Not a Bird?" *Man* 2: 5–25.

Bulmer, Ralph. 1982. "Traditional Conservation Practices in Papua New Guinea." In *Traditional Conservation in Papua New Guinea: Implications for Today*, ed. L. Morauta, J. Pernetta, and W. Heaney, 59–77. Boroko, PNG: Institute of Applied Social and Economic Research.

Burton, John. 1991. *Porgera Census Project: Report for 1990*. Port Moresby: Department of Anthropology and Sociology, University of Papua New Guinea.

Carpenter, Steve, et al. 2001. "From Metaphor to Measurement: Resilience of What to What." *Ecosystems* 4 (8): 765–81.

Casey, Edward S. 1996. "How to Get from Space to Place in a Fairly Short Stretch of Time." In *Senses of Place*, ed. S. Feld and K. H. Basso, 13–52. Santa Fe, NM: School of American Research Press.

Castree, Noel. 1995. "The Nature of Produced Nature: Materiality and Knowledge Construction in Marxism." *Antipode* 27 (1): 12–48.

Chapin, F. Stuart, III, Carl Folke, and Gary Kofinas. 2009. "A Framework for Understanding Change." In *Principles of Ecosystem Stewardship: Resilience-Based Natural Resource Management in a Changing World*, ed. F. S. Chapin III, G. Kofinas, and C. Folke, 3–28. New York: Springer.

Chapman, Colin, and Lauren Chapman. 2004. "Unfavorable Successional Pathways and the Conservation Value of Logged Tropical Forest." *Biodiversity and Conservation* 13: 2089–105.

Clark, Jeffrey. 1993. "Gold, Sex, and Pollution: Male Illness and Myth at Mt. Kare, Papua New Guinea." *American Ethnologist* 20 (4): 742–57.

Clark, Jeffrey. 1995. "Highlands of History: Images of Deviance and Desire." In *Papuan Borderlands: Huli, Duna, and Ipili Perspectives on the Papua New Guinea Highlands*, ed. A. Biersack, 379–400. Ann Arbor: University of Michigan Press.

Clark, John. 1944. "Report on Twelve Months Activities, Wabag Sub-District." Unpublished manuscript.

Clarke, William. 1971. *Place and People: An Ecology of a New Guinea Community*. Berkeley: University of California Press.

Clements, Frederic E. 1916. *Plant Succession: An Analysis of the Development of Vegetation*. Washington, DC: Carnegie Institute of Washington.

Conklin, Harold. 1954. "An Ethnoecological Approach to Shifting Agriculture." *Transactions of the New York Academy of Sciences* 17: 133–42.

Connell, John. 1997. *Papua New Guinea: The Struggle for Development*. London: Routledge.

Connell, John, and Richard Howitt, eds. 1991. *Mining and Indigenous Peoples in Australasia*. Sydney: University of Sydney Press.

Connolly, Bob, and Robin Anderson. 1987. *First Contact: New Guinea's Highlanders Encounter the Outside World*. New York: Viking.

Coumans, Catherine. 2011. "Occupying Spaces Created by Conflict: Anthropologists, Development NGOs, Responsible Investment, and Mining." *Current Anthropology* 52 (S3): S29–S43.

Cruikshank, Julie. 2005. *Do Glaciers Listen? Local Knowledge, Colonial Encounters, and Social Imagination*. Vancouver: University of British Columbia Press.

Cruikshank, Julie. 2012. "Are Glaciers 'Good to Think With'? Recognising Indigenous Environmental Knowledge." *Anthropological Forum* 22 (3): 239–50.

Crumley, Carole, ed. 1994. *Historical Ecology: Cultural Knowledge and Changing Landscapes*. Santa Fe, NM: School of American Research Press.

Crush, Jonathan, ed. 1995. *Power of Development*. New York: Routledge.

Curran, L. M., et al. 2004. "Lowland Forest Loss in Protected Areas of Indonesian Borneo." *Science* 303: 1000–1003.

Derkley, Harry, ed. 1989. *The Porgera Agreements*. Madang, PNG: Enga Provincial Government.

Descola, Philippe. 2013. *Beyond Nature and Culture*. Trans. J. Lloyd. Chicago: University of Chicago Press.

Descola, Philippe, and G. Pálsson, eds. 1996. *Nature and Society: Anthropological Perspectives*. New York: Routledge.

Dirlik, Arif. 2001. "Place-based Imagination: Globalism and the Politics of Place." In *Places and Politics in an Age of Globalization*, ed. R. Prazniak and A. Dirlik, 15–51. Lanham, MD: Rowman and Littlefield.

Dixon, Peter B., Gae Kauzi, and Maureen T. Rimmer. 2010. "Effects on the PNG Economy of a Major LNG Project." *Economic Papers* 29 (2): 143–55.

Dove, Michael R. 2006. "Equilibrium Theory and Interdisciplinary Borrowing: A Comparison of Old and New Ecological Anthropologies." In *Reimagining Political Ecology*, ed. A. Biersack and J. B. Greenberg, 43–69. Durham, NC: Duke University Press.

Dove, Michael R. 2011. *The Banana Tree at the Gate: A History of Marginal Peoples and Global Markets in Borneo*. New Haven, CT: Yale University Press.

Dove, Michael R., and D. Kammen. 1997. "The Epistemology of Sustainable Resource Use: Managing Forest Products, Swiddens, and High-Yielding Variety Crops." *Human Organization* 56 (1): 91–101.

Dundon, Alison. 2002. "Mines and Monsters: A Dialogue on Development in Western Province, Papua New Guinea." *Australian Journal of Anthropology* 13 (2): 139–54.

Durkheim, Émile. 1997. *Suicide: A Study in Sociology*. New York: Free Press.

Dwyer, Peter D., and Monica Minnegal. 2010. "Theorizing Social Change." *Journal of the Royal Anthropological Institute* 16 (3): 629–45.

Ernst, Thomas M. 1999. "Land, Stories, and Resources: Discourse and Entification in Onbasulu Modernity." *American Anthropologist* 101: 88–97.

Errington, Frederick, and Deborah Gewertz. 2004. *Yali's Question: Sugar, Culture and History*. Chicago: University of Chicago Press.

Escobar, Arturo. 1995. *Encountering Development: The Making and Unmaking of the Third World*. Princeton, NJ: Princeton University Press.

Escobar, Arturo. 1999. "After Nature: Steps to an Anti-essentialist Political Ecology." *Current Anthropology* 40: 1–30.

Escobar, Arturo. 2001. "Culture Sits in Places: Reflections on Globalism and Subaltern Strategies of Localization." *Political Geography* 20: 139–74.

Escobar, Arturo. 2008. *Territories of Difference: Place, Movements, Life, Redes*. Durham, NC: Duke University Press.

Fairhead, James, Melissa Leach, and Ian Scoones. 2012. "Green Grabbing: A New Appropriation of Nature?" *Journal of Peasant Studies* 39 (2): 237–61.

FAO. 2010. *Global Forest Resources Assessment 2010.* Rome: Food and Agriculture Organization of the United Nations.

Feil, Daryl Keith. 1983. "World without Exchange: Millenia and the Tee Ceremonial System in Tombema-Enga Society." *Anthropos* 78: 89–106.

Feld, Steven, and Keith H. Basso. 1996. Introduction to *Senses of Place*, ed. S. Feld and K. H. Basso, 3–11. Santa Fe, NM: School of American Research Press.

Ferguson, James. 1994. *The Anti-Politics Machine: "Development," Depoliticization, and Bureaucratic Power in Lesotho.* Minneapolis: University of Minnesota Press.

Ferguson, James. 1999. *Expectations of Modernity: Myths and Meanings of Urban Life on the Zambian Copperbelt.* Berkeley: University of California Press.

Ferguson, James. 2006. *Global Shadows: Africa in the Neoliberal World Order.* Durham, NC: Duke University Press.

Filer, Colin. 1990. "The Bougainville Rebellion, the Mining Industry and the Process of Social Disintegration in Papua New Guinea." In *The Bougainville Crisis*, ed. R. J. May and M. Spriggs, 73–112. Bathurst, Australia: Crawford House Press.

Filer, Colin. 1997. "Compensation, Rent and Power in Papua New Guinea." In *Compensation for Resource Development in Papua New Guinea*, ed. S. Toft, 156–90. Port Moresby: Law Reform Commission of Papua New Guinea.

Filer, Colin. 1999. *Dilemmas of Development: The Social and Economic Impact of the Porgera Gold Mine, 1989–1994.* Boroko, PNG: National Research Institute.

Filer, Colin, and Martha MacIntyre. 2006. "Grass Roots and Deep Holes: Community Responses to Mining in Melanesia." *Contemporary Pacific* 18 (2): 215–31.

Filer, Colin, and Michael Wood. 2012. "The Creation and Dissolution of Private Property in Forest Carbon: A Case Study from Papua New Guinea." *Human Ecology* 40 (5): 665–77.

Finlayson, Mike. 2002. "Benefit Stream Analysis: Working Paper 2." Canberra. Submitted to the Government of Papua New Guinea under the PNG Mining Sector Institutional Strengthening Project.

Flannery, Tim. 1995. *Mammals of New Guinea.* Chatswood, Australia: Reed Books.

Foster, John Bellamy. 2000. *Marx's Ecology: Materialism and Nature.* New York: Monthly Review Press.

Foster, John Bellamy, Brett Clark, and Richard York. 2010. *The Ecological Rift: Capitalism's War on the Earth.* New York: Monthly Review Press.

Fox, T. A., and J. Fox. 1936. "In New Guinea's Unknown Centre." *Pacific Islands Monthly* (February): 41–44.

Frankel, Stephen. 1986. *The Huli Response to Illness.* Cambridge: Cambridge University Press.

Friedman, Jonathan. 1974. "Marxism, Structuralism, and Vulgar Materialism." *Man* 9: 444–69.

Gammage, Bill. 1998. *The Sky Travellers: Journeys in New Guinea, 1938–1939.* Victoria: Melbourne University Press.

Geertz, Clifford. 1973. *The Interpretation of Cultures: Selected Essays*. New York: Basic Books.

Gewertz, Deborah, and Frederick Errington. 1999. *Emerging Class in Papua New Guinea: The Telling of Difference*. New York: Cambridge University Press.

Gibbs, Philip. 1975. "Ipili Religion Past and Present." MA thesis, University of Sydney.

Gibbs, Philip. 1977. "The Cult from Lyeimi and the Ipili." *Oceania* 48 (1): 1–25.

Gilberthorpe, Emma. 2007. "Fasu Solidarity: A Case Study of Kin Networks, Land Tenure, and Oil Extraction in Kutubu, Papua New Guinea." *American Anthropologist* 109 (1): 101–12.

Gimpel, Stephen. 2008. "Illegal Mining and Sustainability: Porgera Joint Venture, PNG." In *The Resources Sector: Securing Sustainable Livelihoods for Communities*. Brisbane, Australia: Foundation for Development Cooperation.

Glasse, Robert M. 1995. "Time Belong *Mbingi*: Religious Syncretism and the Pacification of the Huli." In *Papuan Borderlands: Huli, Duna, and Ipili Perspectives on the Papua New Guinea Highlands*, ed. A. Biersack, 57–86. Ann Arbor: University of Michigan Press.

Goldman, Laurence. 1983. *Talk Never Dies: The Language of Huli Disputes*. London: Tavistock.

Goldman, Laurence, and Chris Ballard, eds. 1998. *Fluid Ontologies: Myth, Ritual and Philosophy in the Highlands of Papua New Guinea*. Westport, CT: Bergin and Garvey.

Golub, Alex. 2006. "Making the Ipili Feasible: Imagining Local and Global Actors at the Porgera Gold Mine, Enga Province, Papua New Guinea." Ph.D. diss., University of Chicago.

Golub, Alex. 2007. "Ironies of Organization: Landowners, Land Registration, and Papua New Guinea's Mining and Petroleum Industry." *Human Organization* 66 (1): 38–48.

Golub, Alex. 2014. *Leviathans at the Gold Mine: Creating Indigenous and Corporate Actors in Papua New Guinea*. Durham, NC: Duke University Press.

Gressitt, J. L. 1982. "General Introduction." In *Biogeography and Ecology of New Guinea*, ed. J. L. Gressitt, 3–13. The Hague: Dr. W. Junk.

Grossman, Lawrence. 1984. *Peasants, Subsistence Ecology, and Development in the Highlands of Papua New Guinea*. Princeton, NJ: Princeton University Press.

Gunderson, Lance, and C. S. Holling, eds. 2002. *Panarchy: Understanding Transformations in Human and Natural Systems*. Washington, DC: Island Press.

Gunson, M. J., et al. 1997. "Stratigraphic Reconstruction of the Porgera Region, Papua New Guinea." In *Proceedings: PNG Geology, Exploration and Mining Conference, 1997, Madang, Papua New Guinea*, ed. G. E. Hancock, 99–108. Carlton, Australia: Australasian Institute of Mining and Metallurgy.

Gunson, Mark, Greg Hall, and Mike Johnston. 2000. "Foraminiferal Coloration Index as a Guide to Hydrothermal Gradients around the Porgera Intrusive Complex, Papua New Guinea." *Economic Geology* 95: 271–82.

Gupta, Akhil, and James Ferguson. 1992. "Beyond 'Culture': Space, Identity, and the Politics of Difference." *Cultural Anthropology* 7: 6–23.

Gupta, Akhil, and James Ferguson. 1997. *Culture, Power, Place: Explorations in Critical Anthropology*. Durham, NC: Duke University Press.

Gurnis, M., R. Mueller, and L. Moresi. 1998. "Dynamics of Cretaceous Vertical Motion of Australia and the Australian-Antarctic Discordance." *Science* 279: 1499–504.

Haenn, Nora. 1999. "The Power of Environmental Knowledge: Ethnoecology and Environmental Conflicts in Mexican Conservation." *Human Ecology* 27 (3): 477–91.

Hagender, F. 2000. *The Spirit of Trees: Science, Symbiosis and Inspiration*. Edinburgh: Floris Books.

Halvaksz, Jamon. 2003. "Singing about the Land among the Biangai." *Oceania* 73 (3): 153–69.

Halvaksz, Jamon. 2006. "Cannibalistic Imaginaries: Mining the Natural and Social Body in Papua New Guinea." *Contemporary Pacific* 18 (2): 335–59.

Halvaksz, Jamon. 2008. "Whose Closure? Appearances, Temporality, and Mineral Extraction in Papua New Guinea." *Journal of the Royal Anthropological Institute* 14: 21–37.

Handley, G., and D. Henry. 1987. "Porgera Environmental Plan: Report on Small Scale Alluvial Mining." Unpublished manuscript, Porgera District Administration Archives, PNG.

Hardesty, Donald L. 1975. "The Niche Concept: Suggestions for Its Use in Human Ecology." *Human Ecology* 3 (2): 71–85.

Hirsch, Eric, and Michael O'Hanlon, eds. 1995. *The Anthropology of Landscape: Perspectives on Place and Space*. Oxford: Oxford University Press.

Holling, C. S. 1973. "Resilience and Stability of Ecological Systems." *Annual Review of Ecology and Systematics* 4: 1–23.

Holling, C. S. 2001. "Understanding the Complexity of Economic, Ecological, and Social Systems." *Ecosystems* 4: 390–405.

Holling, C. S., and Lance Gunderson. 2002. "Resilience and Adaptive Cycles." In *Panarchy: Understanding Transformations in Human and Natural Systems*, ed. L. Gunderson and C. S. Holling, 25–62. Washington, DC: Island Press.

Holling, C. S., Lance Gunderson, and Donald Ludwig. 2002a. "In Quest of a Theory of Adaptive Change." In *Panarchy: Understanding Transformations in Human and Natural Systems*, ed. L. Gunderson and C. S. Holling, 3–22. Washington, DC: Island Press.

Holling, C. S., Lance Gunderson, and Garry D. Peterson. 2002b. "Sustainability and Panarchies." In *Panarchy: Understanding Transformations in Human and Natural Systems*, ed. L. Gunderson and C. S. Holling, 63–102. Washington, DC: Island Press.

Hutchinson, G. E. 1965. *The Ecological Theater and the Evolutionary Play*. New Haven, CT: Yale University Press.

Hviding, Edvard. 1996. *Guardians of Marovo Lagoon: Practice, Place, and Politics in Maritime Melanesia*. Honolulu: University of Hawai'i Press.

Hviding, Edvard, and Tim Bayliss-Smith. 2000. *Islands of Rainforest: Agroforestry, Logging, and Eco-tourism in the Solomon Islands.* Burlington, VT: Ashgate.

Hyndman, David. 1984. "Ethnobotany of Wopkaimin *Pandanus*: Significant Papua New Guinea Plant Resource." *Economic Botany* 38 (3): 287–303.

Hyndman, David. 1994. *Ancestral Rain Forests and the Mountain of Gold: Indigenous Peoples and Mining in New Guinea.* Boulder, CO: Westview Press.

Imbun, B., and P. McGavin, eds. 2001. *Mining in Papua New Guinea: Analysis and Policy Implications.* Waigani: University of Papua New Guinea Press.

Ingemann, Frances. n.d. "Ipili Dictionary." Unpublished manuscript.

Ingold, Tim. 2000. *The Perception of the Environment: Essays in Livelihood, Dwelling, and Skill.* London: Routledge.

Jacka, Jerry K. 2001. "Coca-Cola and *Kolo*: Land, Ancestors, and Development." *Anthropology Today* 17 (4): 3–8.

Jacka, Jerry K. 2002. "Cults and Christianity among the Enga and Ipili." *Oceania* 72 (3): 196–214.

Jacka, Jerry K. 2003. "God, Gold, and the Ground: Place-Based Political Ecology in a New Guinea Borderlands." PhD diss., University of Oregon.

Jacka, Jerry K. 2005. "Emplacement and Millennial Expectations in an Era of Development and Globalization: Heaven and the Appeal of Christianity among the Ipili." *American Anthropologist* 107 (4): 643–53.

Jacka, Jerry K. 2007. "Whitemen, the Ipili and the City of Gold: A History of the Politics of Race and Development in Highlands New Guinea." *Ethnohistory* 54 (3): 445–71.

Jacka, Jerry K. 2009. "Global Averages, Local Extremes: The Subtleties and Complexities of Climate Change in Papua New Guinea." In *Anthropology and Climate Change: From Encounters to Actions*, ed. S. Crate and M. Nutall, 197–208. Walnut Grove, CA: Left Coast Press.

Jacka, Jerry K. 2010. "The Spirits of Conservation: Ecology, Christianity, and Resource Management in Highlands Papua New Guinea." *Journal for the Study of Religion, Nature and Culture* 4 (1): 24–47.

Jackson, Richard, and Glenn Banks. 2002. *In Search of the Serpent's Skin: The Story of the Porgera Gold Project.* Port Moresby, PNG: Placer Niugini.

Johns, R. J. 1982. "Plant Zonation." In *Biogeography and Ecology of New Guinea*, ed. J. L. Gressitt, 1: 309–30. The Hague: Dr. W. Junk.

Johnson, Penny. 2011. *Scoping Project: Social Impact of the Mining Project on Women in the Porgera Area.* Port Moresby, PNG: Porgera Environmental Advisory Komiti.

Johnson, Peter. 2012. *Lode Shedding: A Case Study of the Economic Benefits to the Landowners, the Provincial Government, and the State, from the Porgera Gold Mine: Background and Financial Flows from the Mine.* Port Moresby, PNG: National Research Institute.

Jones, Owain, and Paul Cloke. 2002. *Tree Cultures: The Place of Trees and Trees in Their Place.* Oxford: Berg.

Jorgensen, Dan. 1990. "Placing the Past and Moving the Present: Myth and Contemporary History in Telefolmin." *Culture* 10: 47–56.

Jorgensen, Dan. 1996. "Regional History and Ethnic Identity in the Hub of New Guinea: The Emergence of the Min." *Oceania* 66 (3): 189–210.

Jorgensen, Dan. 1998. "Whose Nature? Invading Bush Spirits, Travelling Ancestors, and Mining in Telefolmin." *Social Analysis* 42 (3): 100–116.

Jorgensen, Dan. 2001. "Who and What Is a Landowner? Mythology and Marking the Ground in a Papua New Guinea Mining Project." In *Mining and Indigenous Lifeworlds in Australia and Papua New Guinea*, ed. A. Rumsey and J. Weiner, 101–24. Wantage, UK: Sean Kingston.

Jorgensen, Dan. 2014. "Mining Narratives and Multiple Geographies in Papua New Guinea: Ok Tedi, the Emerald Cave and Lost Tribes." *Journal de la Société des Océanistes* 138: 23–36.

Kalb, Don. 2013. "Financialization and the Capitalist Moment: Marx versus Weber in the Anthropology of Global Systems." *American Ethnologist* 40 (2): 258–66.

Kirsch, Stuart. 2001. "Lost Worlds: Environmental Disaster, 'Culture Loss,' and the Law." *Current Anthropology* 42 (2): 167–98.

Kirsch, Stuart. 2006. *Reverse Anthropology: Indigenous Analysis of Social and Environmental Relations in New Guinea*. Stanford, CA: Stanford University Press.

Kirsch, Stuart. 2008. "Social Relations and the Green Critique of Capitalism in Melanesia." *American Anthropologist* 110 (3): 288–98.

Kirsch, Stuart. 2014. *Mining Capitalism: The Relationship between Corporations and Their Critics*. Berkeley: University of California Press.

Kituai, August. 1998. *My Gun, My Brother: The World of the Papua New Guinea Colonial Police, 1920–1960*. Honolulu: University of Hawai'i Press.

Knauft, Bruce. 2002. *Exchanging the Past: A Rainforest World of Before and After*. Chicago: University of Chicago Press.

Kohn, Eduardo. 2007. "How Dogs Dream: Amazonian Natures and the Politics of Transspecies Engagement." *American Ethnologist* 34 (1): 3–24.

Krebs, Charles. 1989. *Ecological Methodology*. New York: HarperCollins.

Kulick, Don. 1992. "Coming Up in Gapun: Conceptions of Development and Their Effect on Language in a Papua New Guinea Village." In *Kam-ap or Take-off: Local Notions of Development*, ed. G. Dahl and A. Rabo, 10–34. Uppsala: Stockholm Studies in Social Anthropology.

Latour, Bruno. 2009. "Perspectivism: 'Type' or 'Bomb.'" *Anthropology Today* 25 (2): 1–2.

Laurance, William F., Miriam Goosem, and Susan G. W. Laurance. 2009. "Impacts of Roads and Linear Clearings on Tropical Forests." *Trends in Ecology and Evolution* 24 (12): 659–69.

Leahy, Michael. 1991. *Explorations into Highland New Guinea 1930–1935*. Tuscaloosa: University of Alabama Press.

Leopold, Aldo. 1949. *A Sand County Almanac*. New York: Oxford University Press.

Leslie, Paul, and J. Terrence McCabe. 2013. "Response Diversity and Resilience in Social-Ecological Systems." *Current Anthropology* 54 (2): 114–44.

Lewis, M. Paul, Gary Simons, and Charles Fennig. 2013. *Ethnologue: Languages of the World*, 17th ed. Dallas: SIL International.

Li, Tanya Murray. 2007. *The Will to Improve: Governmentality, Development, and the Practice of Politics*. Durham, NC: Duke University Press.

Linden, Stanton, ed. 2003. *The Alchemy Reader: From Hermes Trismegistus to Isaac Newton*. Cambridge: Cambridge University Press.

Löffler, Ernst. 1977. *Geomorphology of Papua New Guinea*. Canberra: CSIRO, Australian National University.

Macintyre, Martha, and Simon Foale. 2004. "Politicised Ecology: Local Responses to Mining in Papua New Guinea." *Oceania* 74 (3): 231–51.

Malinowski, Bronislaw. 1984. *Argonauts of the Western Pacific*. Prospect Heights, IL: Waveland Press.

Manner, H. I. 1981. "Ecological Succession in New and Old Swiddens of Montane Papua New Guinea." *Human Ecology* 9 (3): 359–77.

Marx, Karl. 1973. *Grundrisse: Foundations of the Critique of Political Economy*. Trans. M. Nicolaus. New York: Random House.

Marx, Karl. 1990. *Capital: A Critique of Political Economy*. Vol. 1. Trans. B. Fowkes. New York: Penguin Books.

Massey, Doreen. 1994. *Space, Place, and Gender*. Minneapolis: University of Minnesota Press.

May, R. J., and Matthew Spriggs, eds. 1990. *The Bougainville Crisis*. Bathurst, Australia: Crawford House Press.

McAlpine, J., and D. F. Freyne. 2001. "Land Use Change and Intensification in Papua New Guinea 1975–1996." *Asia Pacific Viewpoint* 42 (2–3): 209–18.

McAlpine, J., Gael Keig, and R. Falls. 1983. *Climate of Papua New Guinea*. Canberra: CSIRO, Australian National University.

Meggitt, Mervyn J. 1956. "The Valleys of the Upper Wage and Lai Rivers, Western Highlands, New Guinea." *Oceania* 27: 90–135.

Meggitt, Mervyn J. 1964. "Male-Female Relationships in the Highlands of Australian New Guinea." *American Anthropologist* 66: 204–24.

Meggitt, Mervyn J. 1973. "The Sun and the Shakers: A Millenarian Cult and Its Transformations in the New Guinea Highlands." *Oceania* 44 (1): 1–56.

Meggitt, Mervyn J. 1974. "'Pigs Are Our Hearts!' The *Te* Exchange Cycle among the Mae Enga of New Guinea." *Oceania* 44: 165–203.

Merchant, Carolyn. 1990. *The Death of Nature: Women, Ecology, and the Scientific Revolution*. New York: HarperOne.

Mintz, Sidney. 1985. *Sweetness and Power: The Place of Sugar in Modern History*. New York: Penguin Books.

Mittermeier, R. A., et al. 2003. "Wilderness and Biodiversity Conservation." *Proceedings of the National Academy of Sciences* 100 (18): 10309–13.

Modjeska, Nicholas. 1982. "Production and Inequality: Perspectives from Central New Guinea." In *Inequality in New Guinea Highlands Societies*, ed. A. Strathern, 50–108. Cambridge: Cambridge University Press.

Moore, Donald S. 1993. "Contesting Terrain in Eastern Zimbabwe's Highlands: Political Ecology, Ethnography, and Peasant Resource Struggles." *Economic Geography* 69 (4): 380–401.

Moran, Bruce. 2005. *Distilling Knowledge: Alchemy, Chemistry, and the Scientific Revolution.* Cambridge, MA: Harvard University Press.

Morphy, Howard. 1990. "Myth, Totemism and the Creation of Clans." *Oceania* 60: 312–28.

Morphy, Howard. 1995. "Landscape and the Reproduction of the Ancestral Past." In *The Anthropology of Landscape: Perspectives on Place and Space*, ed. E. Hirsch and M. O'Hanlon, 184–209. Oxford: Clarendon Press.

Munn, Nancy. 1986. *The Fame of Gawa: A Symbolic Study of Value Transformation in a Massim (Papua New Guinea) Society.* Cambridge: Cambridge University Press.

Nadasdy, Paul. 2007. "The Gift in the Animal: The Ontology of Hunting and Human-Animal Sociality." *American Ethnologist* 34 (1): 25–43.

Nazarea, Virginia D. 1999. "A View from a Point: Ethnoecology as Situated Knowledge." In *Ethnoecology: Situated Knowledge/Located Lives*, ed. V. D. Nazarea, 3–20. Tucson: University of Arizona Press.

Nelson, Donald R., W. Neil Adger, and Katrina Brown. 2007. "Adaptation to Environmental Change: Contributions of a Resilience Framework." *Annual Review of Environment and Resources* 32: 395–419.

Newman, William R. 2005. *Promethean Ambitions: Alchemy and the Quest to Perfect Nature.* Chicago: University of Chicago Press.

Nita, Albert. 2001. "New Power Structures and Environmental Management: Evidence from Porgera Gold Mine." In *Mining in Papua New Guinea: Analysis and Policy Implications*, ed. B. Imbun and P. McGavin, 157–72. Waigani: University of Papua New Guinea Press.

Norberg, Jon, and Graeme S. Cumming, eds. 2008. *Complexity Theory for a Sustainable Future.* New York: Columbia University Press.

Oates, John. 1999. *Myth and Reality in the Rain Forest: How Conservation Strategies Are Failing in West Africa.* Berkeley: University of California Press.

Odum, Eugene. 1969. "The Strategy of Ecosystem Development." *Science* 164 (3877): 262–70.

Ogden, Laura, et al. 2013. "Global Assemblages, Resilience, and Earth Stewardship in the Anthropocene." *Frontiers in Ecology and Environment* 11 (7): 341–47.

Ong, Aihwa. 2010. *Spirits of Resistance and Capitalist Discipline: Factory Women in Malaysia.* Albany: State University of New York Press.

Orlove, Ben. 1980. "Ecological Anthropology." *Annual Review of Anthropology* 9: 235–73.

Ostrom, Elinor. 1990. *Governing the Commons: The Evolution of Institutions for Collective Action.* Cambridge: Cambridge University Press.

Ovid. 2010. *Metamorphoses.* Trans. S. Lombardo. Indianapolis: Hackett.

Pacific Agribusiness. 1987. *Social and Economic Impact Study: Porgera Gold Mine.* 2 vols. Port Moresby, PNG: Department of Minerals and Energy.

Paijmans, K. 1976. *New Guinea Vegetation.* Amsterdam: Elsevier Scientific.

Paulson, Susan, and Lisa L. Gezon, eds. 2005. *Political Ecology across Spaces, Scales, and Social Groups.* New Brunswick, NJ: Rutgers University Press.

Paulson, Susan, Lisa L. Gezon, and M. Watts. 2005. "Politics, Ecologies, Genealogies." In *Political Ecology across Spaces, Scales, and Social Groups*, ed. S. Paulson and L. L. Gezon, 17–37. New Brunswick, NJ: Rutgers University Press.

Peet, Richard, and Michael Watts. 1993. "Introduction: Development Theory and Environment in an Age of Market Triumphalism." *Economic Geography* 69: 227–53.

Peluso, Nancy, and Christian Lund. 2011. "New Frontiers of Land Control: Introduction." *Journal of Peasant Studies* 38 (4): 667–81.

Peters, F. E., Monette Tomono, and Pamela A. Wills. 1960. "Composition of Some South Pacific Foods." *Journal of Food Science* 25 (2): 211–28.

Primack, R., and R. Corlett. 2005. *Tropical Rain Forests: An Ecological and Biogeographical Comparison.* Oxford: Blackwell.

Puge, Jeffrey. 1992. Letter from Acting District Manager, Porgera, to the Secretary, Department of Enga. October 20. Porgera District Administration Archives. Porgera, Papua New Guinea.

Rappaport, Roy. 1967. "Ritual Regulation of Environmental Relations among a New Guinea People." *Ethnology* 6: 17–30.

Rappaport, Roy. 1971. "The Flow of Energy in an Agricultural System." *Scientific American* 25: 116–32.

Rappaport, Roy. 1984. *Pigs for the Ancestors: Ritual in the Ecology of a New Guinea People.* New Haven, CT: Yale University Press.

Rappaport, Roy. 1993. "Distinguished Lecture in General Anthropology: The Anthropology of Trouble." *American Anthropologist* 95 (2): 295–303.

Reed, Kristin. 2009. *Crude Existence: Environment and the Politics of Oil in Northern Angola.* Berkeley: University of California Press.

Reichel-Dolmatoff, Gerardo. 1996. *The Forest Within: The World-view of the Tukano Amazonian Indians.* Totnes, UK: Themis.

Richards, J., et al. 1997. "Fluid Chemistry and Processes at the Porgera Gold Deposit, Papua New Guinea." *Mineralium Deposita* 32: 119–32.

Rist, Gilbert. 2006. *The History of Development: From Western Origins to Global Faith.* New York: Zed Books.

Rival, Laura, ed. 1998. *The Social Life of Trees: Anthropological Perspectives on Tree Symbolism.* Oxford: Berg.

Robbins, Joel. 2004. *Becoming Sinners: Christianity and Moral Torment in a Papua New Guinea Society.* Berkeley: University of California Press.

Robbins, Joel, and Holly Wardlow, eds. 2005. *The Making of Global and Local Modernities in Melanesia: Humiliation, Transformation, and the Nature of Culture Change.* Burlington, VT: Ashgate.

Robbins, Paul. 2004. *Political Ecology: A Critical Introduction.* Oxford: Blackwell.

Robinson, N. C. 1960. "Report of Extended Patrol in the Native Mining Area of the Porgera River Western Highlands, August–October 1960." Unpublished manuscript.

Rodman, Margaret. 1992. "Empowering Place: Multilocality and Multivocality." *American Anthropologist* 94: 640–56.

Ronacher, E., J. Richards, and M. Johnston. 2000. "Evidence for Fluid Phase Separation in High-Grade Ore Zones at the Porgera Gold Deposit, Papua New Guinea." *Mineralium Deposita* 35: 683–88.

Rumsey, Alan, and James Weiner, eds. 2001. *Emplaced Myth: Space, Narrative, and Knowledge in Aboriginal Australia and Papua New Guinea.* Honolulu: University of Hawai'i Press.

Rumsey, Alan, and James Weiner. 2004. *Mining and Indigenous Lifeworlds in Australia and Papua New Guinea.* Wantage, UK: Sean Kingston.

Ryan, Peter. 1991. *Black Bonanza: A Landslide of Gold.* Victoria, Australia: Hyland House.

Sachs, Wolfgang, ed. 1992. *The Development Dictionary: A Guide to Knowledge as Power.* London: Zed Books.

Sahlins, Marshall. 1963. "Poor Man, Rich Man, Big-Man, Chief: Political Types in Melanesia and Polynesia." *Comparative Studies in Society and History* 5: 285–303.

Sahlins, Marshall. 1976. *Culture and Practical Reason.* Chicago: University of Chicago Press.

Sahlins, Marshall. 1981. *Historical Metaphors and Mythical Realities: Structure in the Early History of the Sandwich Islands.* Ann Arbor: University of Michigan Press.

Sahlins, Marshall. 1988. "Cosmologies of Capitalism: The Trans-Pacific Sector of 'The World System.'" *Proceedings of the British Academy* 74: 1–51.

Sahlins, Marshall. 1992. "The Economics of Develop-man in the Pacific." *Res* 21: 12–25.

Sahlins, Marshall. 1994. "Cosmologies of Capitalism: The Trans-Pacific Sector of 'The World System.'" In *Culture/Power/History: A Reader in Contemporary Social Theory,* ed. N. Dirks, G. Eley, and S. Ortner, 412–55. Princeton, NJ: Princeton University Press.

Sahlins, Marshall. 1999. "What Is Anthropological Enlightenment? Some Lessons of the Twentieth Century." *Annual Review of Anthropology* 28: i–xxiii.

Scheffer, Martin, et al. 2002. "Dynamic Interaction of Societies and Ecosystems-Linking Theories from Ecology, Economy, and Sociology." In *Panarchy: Understanding Transformations in Human and Natural Systems,* ed. L. Gunderson and C. S. Holling, 195–239. Washington, DC: Island Press.

Schieffelin, Edward. 1991. "The Historical Impact: The Reverberations of First Contact in Subsequent History." In *Like People You See in a Dream: First Contact in Six Papuan Societies,* ed. E. Schieffelin and R. Crittenden, 283–92. Stanford, CA: Stanford University Press.

Scoones, Ian. 1999. "New Ecology and the Social Sciences: What Prospects for a Fruitful Engagement?" *Annual Review of Anthropology* 28: 479–507.

Scott, James. 1985. *Weapons of the Weak: Everyday Forms of Peasant Resistance.* New Haven, CT: Yale University Press.

Scott, James. 1998. *Seeing Like a State: How Certain Schemes to Improve the Human Condition Have Failed.* New Haven, CT: Yale University Press.

Sharp, Peter. 1990. "The Searching Sun: The Lyeime Movement—Crisis, Tragic Events and Folie à Deux in the Papua New Guinea Highlands." *Papua New Guinea Medical Journal* 33: 111–20.

Shearman, Phil. 2001. "Giving Away Another River: An Analysis of the Impacts of the Porgera Mine on the Strickland River System." In *Mining in Papua New Guinea: Analysis and Policy Implications*, ed. B. Imbun and P. McGavin, 173–90. Port Moresby: University of Papua New Guinea Press.

Shearman, Phil, et al. 2008. *The State of the Forests of Papua New Guinea: Mapping the Extent and Condition of Forest Cover and Measuring the Drivers of Forest Change in the Period 1972–2002*. Port Moresby: University of Papua New Guinea.

Sillitoe, Paul. 1983. *Roots of the Earth: Crops in the Highlands of Papua New Guinea*. Kensington: New South Wales University Press.

Sillitoe, Paul. 1996. *A Place against Time: Land and Environment in the Papua New Guinea Highlands*. Amsterdam: Harwood Academic.

Sillitoe, Paul. 1999. "Beating the Boundaries: Land Tenure and Identity in the Papua New Guinea Highlands." *Journal of Anthropological Research* 55 (3): 331–60.

Sillitoe, Paul. 2010. *From Land to Mouth: The Agricultural "Economy" of the Wola of the New Guinea Highlands*. New Haven, CT: Yale University Press.

Sillitoe, Paul, and Andrew Strathern, eds. 2002. *Horticulture in Papua New Guinea: Case Studies from the Southern and Western Highlands*. Pittsburgh: University of Pittsburgh Press.

Sivaramakrishnan, K., and Arun Agrawal. 2003. "Regional Modernities in Stories and Practices of Development." In *Regional Modernities: The Cultural Politics of Development in India*, ed. K. Sivaramakrishnan and A. Agrawal, 1–61. Stanford, CA: Stanford University Press.

Smith, Neil. 1996. "The Production of Nature." In *Future Natural: Nature, Science, Culture*, ed. G. Robertson, 35–54. New York: Routledge.

Smith, Neil. 2008. *Uneven Development: Nature, Capital, and the Production of Space*. Athens: University of Georgia Press.

Smith, Pamela. 1997. *The Business of Alchemy: Science and Culture in the Holy Roman Empire*. Princeton, NJ: Princeton University Press.

Stasch, Rupert. 2009. *Society of Others: Kinship and Mourning in a West Papuan Place*. Berkeley: University of California Press.

Steward, Julian. 1955. "The Concept and Method of Cultural Ecology." In *Theory of Culture Change: The Methodology of Multilinear Evolution*, ed. J. Steward, 30–42. Urbana: University of Illinois Press.

Stewart, Pamela, and Andrew Strathern, eds. 1997. *Millennial Markers*. Townsville: Centre for Pacific Studies, James Cook University of North Queensland.

Stewart, Pamela, and Andrew Strathern. 2002. *Remaking the World: Myth, Mining, and Ritual Change among the Duna of Papua New Guinea*. Washington, DC: Smithsonian Institution Press.

Stone, Benjamin C. 1982. "New Guinea Pandanaceae: First Approach to Ecology and Biogeography." In *Biogeography and Ecology of New Guinea*, ed. J. L. Gressitt, 1: 401–36. The Hague: Dr. W. Junk.

Strathern, Andrew. 1971. *The Rope of Moka: Big-men and Ceremonial Exchange in Mount Hagen, New Guinea*. Cambridge: Cambridge University Press.

Strathern, Marilyn. 1980. "No Nature, No Culture: The Hagen Case." In *Nature, Culture and Gender*, ed. C. P. MacCormack and M. Strathern, 174–222. Cambridge: Cambridge University Press.

Strathern, Marilyn. 1985. "Discovering 'Social Control.'" *Journal of Law and Society* 12 (2): 111–34.

Strathern, Marilyn. 1988. *The Gender of the Gift: Problems with Women and Problems with Society in Melanesia*. Berkeley: University of California Press.

Strathern, Marilyn. 1995. *Women in Between: Female Roles in a Male World, Mount Hagen, New Guinea*. Lanham, MD: Rowman and Littlefield.

Struhsaker, Thomas T. 1997. *Ecology of an African Rain Forest: Logging in Kibale and the Conflict between Conservation and Exploitation*. Gainesville: University Press of Florida.

Summerhayes, Glenn, et al. 2010. "Human Adaptation and Plant Use in Highland New Guinea 49,000 to 44,000 Years Ago." *Science* 330 (78): 78–81.

Taylor, Graham. 1993. Letter from First Assistant Secretary Zone 3 Co-ordinator to the Police Commissioner, National Capital District. July 30. Porgera District Administration Archives. Porgera, Papua New Guinea.

Taylor, Jim. 1939. "Hagen-Sepik Patrol 1938–39." In *Hagen-Sepik Patrol Report*, mss 443, Mandeville Special Collections Library, University of San Diego Library.

Togolo, Mel. 2006. "The Resource Curse and Governance: A Papua New Guinea Perspective." In *Globalisation and Governance in the Pacific Islands*, ed. S. Firth, 275–85. Canberra: ANU E Press, Australian National University.

Tsing, Anna. 1993. *In the Realm of the Diamond Queen: Marginality in an Out-of-the-Way Place*. Princeton, NJ: Princeton University Press.

Tsing, Anna. 2000. "Inside the Economy of Appearances." *Public Culture* 12 (1): 115–44.

Tsing, Anna. 2005. *Friction: An Ethnography of Global Connections*. Princeton, NJ: Princeton University Press.

Upton, Phaedra, and Paul Gow. 1998. "Coupled Mechanical/Fluid Flow Modelling of the Papuan Fold Belt Deformation and Fluid Flow." Paper presented at AGC-14 Conference, Townsville, Australia, July 6–10.

Vail, John. 1995. "All That Glitters: The Mt. Kare Gold Rush and Its Aftermath." In *Papuan Borderlands: Huli, Duna, and Ipili Perspectives on the Papua New Guinea Highlands*, ed. A. Biersack, 343–74. Ann Arbor: University of Michigan Press.

van Valkenberg, Johan, and Pieter Ketner. 1994. "Vegetation Changes Following Human Disturbance of Mid-Montane Forest in the Wau Area, Papua New Guinea." *Journal of Tropical Ecology* 10 (1): 41–54.

Vayda, A. P., and B. Walters. 1999. "Against Political Ecology." *Human Ecology* 27: 167–79.

Vitousek, Peter M., et al. 1997. "Human Domination of Earth's Ecosystems." *Science* 277: 494–99.

Viveiros de Castro, Eduardo. 1998. "Cosmological Deixis and Amerindian Perspectivism." *Journal of the Royal Anthropological Institute* 4 (3): 469–88.

Viveiros de Castro, Eduardo. 2004. "Exchanging Perspectives: The Transformation of Objects into Subjects in Amerindian Ontologies." *Common Knowledge* 10 (3): 463–84.

Vogt, Craig. 2012. *International Assessment of Marine and Riverine Disposal of Mine Tailings.* London: United Nations Environment Programme.

Waddell, Eric. 1972. *The Mound Builders: Agricultural Practices, Environment, and Society in the Central Highlands of New Guinea.* Seattle: University of Washington Press.

Waddell, Eric. 1975. "How the Enga Cope with Frost: Responses to Climatic Perturbations in the Central Highlands of New Guinea." *Human Ecology* 3 (4): 249–73.

Wagner, Roy. 1975. *The Invention of Culture.* Chicago: University of Chicago Press.

Walker, Peter A. 2005. "Political Ecology: Where Is the Ecology?" *Progress in Human Geography* 29 (1): 73–82.

Wardlow, Holly. 2001. "The Mount Kare Python: Huli Myths and Gendered Fantasies of Agency." In *Mining and Indigenous Lifeworlds in Australia and Papua New Guinea*, ed. A. Rumsey and J. Weiner, 31–67. Wantage, UK: Sean Kingston.

Wardlow, Holly. 2004. "Anger, Economy, and Female Agency: Problematizing 'Prostitution' and 'Sex Work' among the Huli of Papua New Guinea." *Signs: Journal of Women in Culture and Society* 29 (4): 1017–40.

Wardlow, Holly. 2006. *Wayward Women: Sexuality and Agency in a New Guinea Society.* Berkeley: University of California Press.

Weber, Max. 1997. *The Protestant Ethic and the Spirit of Capitalism.* Trans. T. Parsons. New York: Routledge.

West, Paige. 2005. "Translation, Value, and Space: Theorizing an Ethnographic and Engaged Environmental Anthropology." *American Anthropologist* 107 (4): 632–42.

West, Paige. 2006a. *Conservation Is Our Government Now: The Politics of Ecology in Papua New Guinea.* Durham, NC: Duke University Press.

West, Paige. 2006b. "Environmental Conservation and Mining: Between Experience and Expectation in the Eastern Highlands of Papua New Guinea." *Contemporary Pacific* 18 (2): 295–313.

West, Paige. 2012. *From Modern Production to Imagined Primitive: The Social World of Coffee from Papua New Guinea.* Durham, NC: Duke University Press.

West, Paige, and James Carrier. 2010. "Introduction: Surroundings, Selves, and Others. The Political Economy of Environment and Identity." *Landscape Research* 34 (2): 157–70.

White, Ben, et al. 2012. "The New Enclosures: Critical Perspectives on Corporate Land Deals." *Journal of Peasant Studies* 39 (3–4): 619–47.

Whitmore, T. C. 1984. *Tropical Rain Forests of the Far East.* 2nd ed. Oxford: Clarendon Press.

Whitmore, T. C. 1989. "Canopy Gaps and the Two Major Groups of Forest Trees." *Ecology* 70 (3): 536–38.

Wiessner, Pauline, and Akii Tumu. 1998. *Historical Vines: Enga Networks of Exchange, Ritual, and Warfare in Papua New Guinea.* Washington, DC: Smithsonian Institution Press.

Wiessner, Polly, and Nitze Pupu. 2012. "Toward Peace: Foreign Arms and Indigenous Institutions in a Papua New Guinea Society." *Science* 337: 1651–54.

Wiessner, Polly, and Akii Tumu. 2001. "Averting the Bush Fire Day: Ain's Cult Revisited." In *Ecology and the Sacred: Engaging the Anthropology of Roy A. Rappaport*, ed. E. Messer and M. Lambek, 300–323. Ann Arbor: University of Michigan Press.

Willerslev, Rane. 2007. *Soul Hunters: Hunting, Animism, and Personhood among the Siberian Yukaghirs*. Berkeley: University of California Press.

Williams, Raymond. 1980. *Problems in Materialism and Culture: Selected Essays.* London: Verso.

Willis, K. J., L. Gillison, and T. M. Brncic. 2004. "How 'Virgin' Is Virgin Rainforest?" *Science* 304: 402–3.

Wohlt, Paul. 1989. "Migration from Yumbisa, 1972–1975." *Mountain Research and Development* 9 (3): 224–34.

Wohlt, Paul B. 1995. "System Integrity and Fringe Adaptations." In *Papuan Borderlands: Huli, Duna, and Ipili Perspectives on the Papua New Guinea Highlands*, ed. A. Biersack, 199–228. Ann Arbor: University of Michigan Press.

Wolf, Eric R. 1982. *Europe and the People without History*. Berkeley: University of California Press.

Wood, Andrew. 1983. "Soil Types and Traditional Soil Management in the Purari Catchment." In *The Purari: Tropical Environment of a High Rainfall River Basin*, ed. T. Petr, 67–86. The Hague: Dr. W. Junk.

Page numbers followed by *f* indicate illustrations.

Casey, Edward: place, space, and time, 37

Castanopsis acuminatissima (pai), 134, 154, 190–98

Casuarina oligodon (yawale), 92, 110–12, 179–80, 189, 194–97

census, 46, 124, 181–82, 245nn8–9; coffee, 212; mobility and, 116; wards in Porgera, 14

Christianity: ancestral spirits and, 152–53; climate change and, 174, 176; cults and, 61; development and, 22; disenchantment of landscape and, 82, 237; ecology and, 5–6; gendered space and, 97; globalization and, 23; "land is ending" and, 6, 20, 240; leadership and, 23, 69–70; missionaries, 18–19, 70–71; ontology and, 1, 4, 39; reciprocity and, 3; Second Coming, 176; social transformation and, 7; soil fertility and, 98; syncretization with indigenous beliefs, 71, 146, 166, 237; "villages" and, 95; as work, 105, 201

clan, 114, 116, 118; land ownership in, 83, 85, 123–24, 242n1; sacred pools of, 152

Clark, Jeffrey: gold and money, 200; myth and reverse histories, 235–36; Taiyundika and Mt. Kare, 91

Clark, John F. (ANGAU Captain), 60

Clements, Frederic: ecological succession, 31, 158

climate, 170–76, 244nn5–6; coffee production and, 212; global climate change, 29, 241n2; human affairs and, 3; migration and, 168; subsistence and, 34, 167

coffee production, 109, 212–15, 218–19

colonial attitudes, 53–54, 63–68

colonialism: ecology and, 7; globalization and, 22; leadership and, 232; mobility and, 23; in Porgera Valley, 61–72; work and, 105

community relations, 232

compensation, 127–28; bridewealth, 113–14; clans and, 124; death, 68, 101, 147, 234; *laita*, 121–23; land, 86, 213; marriage, 114, 117–18; mining, 2, 15, 46–47, 73–76, 208, 210, 221–22, 225, 243n5; *pamuk pasin* and, 205; warfare, 117, 150, 154, 220, 222. *See also* exchange

complexity: resilience and, 31–32, 160–61

conflict: alluvial mining and, 63–66; colonial agents and Porgerans, 53, 55–57, 60; development, 9, 232; guns and, 223–24, 234; income inequality and, 2, 199; land and, 4, 202–3; mobility and, 157; Porgera Gold Mine related, 7, 15, 27–28, 219–28; reforestation and, 184; resilience of, 234; resource, 22, 180, 198; simplification and, 161; *tandini* and, 153–55. *See also* guns; killings; warfare

conservation: adaptive cycle and, 158–60; development and, 9–10, 18, 29, 31, 231; ritual and, 35

corporate social responsibility, 232

cosmopolitanism, 203–4

Crater Mountain (wildlife management area), 9, 231

creative destruction, 135, 158, 240

crisis: in social-ecological systems, 34, 59, 163, 166

Cruikshank, Julie: social memory, 163

cultural ecology, 34–35

culture: change, 3–4, 6–8, 31–33, 235–37; money and, 201, 209; nature and, 35, 37, 77–78, 118, 155; ontology and, 38–39; Western, perceptions of, 101

dakglaskaman (men who drive around in cars with tinted windows), 210–11

death, 82, 95, 147, 151–55, 226, 234

deforestation, 182–89; mining related, 7; in Papua New Guinea, 29; political ecology of, 4–5

Department of Primary Industries, 212

Des, Epe, xi, 10, 13–17; cassowary discussion, 98–99; gardening with, 41; *lakisia* discussion, 223–24; LLG experiences, 217–18; Porgera mine as a garden, 25–26

Descola, Philippe: ontological systems, 38–39

development: alchemy and, 7–8, 28, 229; ancestral spirits and, 137–41, 154; benefits, 220–21, 233; climate change and, 174, 176; coffee and, 212–15, 218–20; colonial efforts at, 66–67; conflicts, 220–28; conservation and, 9–10, 18, 29, 31, 231; critiques of, 2–3, 7, 229–30; discourses of, 22; as failed modernity, 230–31;

development (*continued*)
landowners and, 73–74; migration and, 46, 96, 163, 178–98, 203; modernity and, 32, 239; ontology and, 4, 237; *pamuk pasin* and, 204–6; in Papua New Guinea, 29–31; of Porgera Gold Mine, 72–76, 210; Porgeran perceptions of, 1–2, 15, 26–28, 71–72, 142–46, 206–8, 215–18, 231–33; qualitative versus quantitative, 232–33; resource degradation and, 5, 197–98, 233; resource frontiers and, 6, 18, 36, 240; vulnerability and, 160–61, 198; wealth and, 200–201. *See also* money

difference: culture and race, 17, 50, 54, 66; ecological, 34, 158–60; 177, 188–98; economic, 68, 203–8; ethnic and language, 46; globalization and, 22; kinship, 116–17, 157; ontology and, 38; production, means of and relations of, 105

dindi gamu (Huli earth renewal ritual), 135–41, 166

Dionysius, 237–38

disturbance, adaptive cycle and, 158–60

diversity measures, 190–92

Division of Mines, 65

dogs: as clan founder, 118; *nembo* and, 148; spirits causing to run away, 94, 104, 151; *wanda yama* and, 150–51

Dove, Michael: moral ecology, 131

dreams: *tandini* and, 142, 147, 153; *yama au* and, 146; *yu angini wanda* in, 100

drought: ENSO and, 59–60, 167–76; livelihoods and, 19; rituals to end, 84, 134–37; social-ecological perception of, 5, 163, 233. *See also* rainfall

Duna (ethnic group), 46, 91

Dwyer, Peter: adaptation versus transformation, 32

earth joints (*yu kimbuni*), 84, 136, 165–66, 240

earth mother women (*yu angini wanda*), 98–104, 150, 155, 237

eating: capitalism and, 101; clan spirits and, 86, 97, 103; gold and, 26, 84, 143–45, 239–40; the land, 111; money and, 204, 212, 240; nature-culture division and, 120; python spirits and, 140–41, 143–45;

reciprocity and, 206; spirits and, 151–55; warfare and, 154; wealth and, 84, 200; whites and, 57, 101; *yama au* and, 146, 150; *yu angini wanda* and, 100

ecological knowledge, 3, 5, 31, 33, 98–99; capitalism and, 155; climate and, 175; place and, 36–37, 79; resilience and, 160, 163, 233–35

ecology: fetishism, 35; forest, 188–98; as an indigenous philosophy, 3, 5–6, 19, 77–79, 130–31, 234; New Guinea, 28–29; Porgera Valley, 39–43, 120; resilience and, 31–34, 157–61; spirits and, 154–55. *See also* cultural ecology; political ecology

economically important trees, 179–80, 190–98

economic development. *See* development

economic inequalities, 200–201, 203–4, 208–12, 220–25. *See also* income; money

Edie Creek (gold rush site), 49, 57

elevation zones, 89, 92*f*; ecological sampling in, 188–98. *See also* *aiyandaka*; *andakama*; *wapi*

El Niño-Southern Oscillation (ENSO), 19, 42, 163, 167–76, 244n2

emotions, 146–48

employment issues, 44, 63, 73, 201–3, 219

Enga (ethnic group), 11, 14, 46, 141; camps along Enga Highway, 189; cults, 59–61; language, 14, 46, 245n11; marriages into Porgera, 125; migrations during ENSO events, 167–72, 176; Mt. Kare gold rush participation, 91; relations with prospectors, 50–51; rituals, 113, 139

Enga Highway, 43–44, 189, 199, 203–8, 226

Enga Province, 39, 90*f*; colonial administration, 62; political dynamics, 215–18; revenues from mining, 73–75, 218; Tax Credit Scheme and, 213; warfare cessation, 234

Enopi, Peter: colonial leadership, 69–70; *laita* discussion, 121

ENSO. *See* El Niño-Southern Oscillation

entropy, 5, 240

environment: agency of, 108; climatic changes, 174–76; landslides as omens, 82–83; local knowledge of, 98–99, 101–2, 171; mining-related changes, 25–26, 30,

myth: gold and, 145–46; development projects and, 139–40; landscape and, 132; resilience and, 234–37; sky people in, 71, 94

National Oceanic and Atmospheric Administration's Climate Prediction Center, 171–72, 244n5
nature, 19, 79; balance of, 177; commodification of, 6, 21; culture and, 36–37, 77, 106, 118, 155; domination of, 105–6; ethnography of, 35–36; ontology and, 38–39
nembo (mind, will, consciousness), 146, 148–51, 155
nena (far away at a lower elevation), 87, 89, 98
Nete (ethnic group), 46
New Guinea Air Wireless Warning Company, 59
New Guinea Goldfields, Ltd. (mining company), 72
New York Times, 230
Nita, Albert: SML landowners as super big men, 209, 218
nonhuman others, 3–4, 38–39, 78–79, 106, 118, 131, 234
nonlandowners, 2, 6, 10, 28, 76; coffee as alternative income, 212–15; differences from landowners, 204–8; household income, 46–47; marriage with landowners, 125–27; money and, 200–202; wealth disparities, 75–76. *See also* landowners
Nothofagus sp. (*tato*), 42, 92, 190–98

Odum, Eugene: ecological homeostasis, 31
Ok Tedi mine, viii, 10, 30, 230–31, 233, 236
omatisia (puberty ritual), 113
ontological dissonance, 4, 10, 39
ontology, 1, 4–5, 19, 33–34, 38–39, 78, 155, 234–37, 240; *ipane* as central to, 97–98
Operation Ipili, 27, 227–28

pai (*Castanopsis acuminatissima*), 134, 154, 190–98
Paiam (clan), 85–86, 152
Paiam, Papua New Guinea, viii, 11, 27, 40, 43–44; colonial census, 181, 245n8; earth renewal rituals performed in, 85–86, 152; mining agreements development of, 75
Paiam Hospital, 149
Paiam Management Company, 86
Paiela Valley, viii, x, 54–55, 125, 212
Pakoa (clan): coffee plantation and, 213; compensation conflicts, 113–14; *ipa ne* of, 152; killings in colonial era, 56–57; *laita*, 122–23; marriage patterns, 125
pamuk pasin (immoral behavior), 204–8
panarchy, 234–35
pandanus (*Pandanus* spp.), 42–43, 93–94, 99, 110–12, 142, 146, 150–51, 167, 174, 178–81, 190–94, 243n1
Panguna mine, Bougainville, 30, 73–74, 210, 233
Papo, Opis (former MP Porgera-Lagaip District), 76, 209, 215–17
Papua New Guinea: climate of, 42, 182; conservation and, 18, 29, 31, 241n2; deforestation in, 29; development issues in, 2, 9, 29–30, 230–33; ecological characteristics of, 28–29; economy of, 29–30, 233; geology of, 39–41; Independence, 72; land area of, 29; military action in Porgera, 27; Mining Act of 1979, 73, 208; population of, 29; simplification and, 161; state involvement in Porgera mine, 2, 14, 72, 209, 232
Paraia, Jonathan (former LLG president in Porgera), 218
pasindia meri (passenger woman), 204–5
patali (surgery, indigenous), 149
patrol officer, 11, 61–62. *See also kiap*
patrols, 62–72, 116, 180–81, 242n4. *See also* Mt. Hagen-Sepik Patrol
Pauwe, John (former LLG president in Porgera), 216–17
PDA (Porgera Development Authority), ix, 27, 74, 216–19, 225; coffee plantation and, 212–14
Pearce, Sir George (colonial minister for territories), 51
Peluso, Nancy: new frontiers of land control, 21
Penale, Ben, xii, 10–12; burning of government compound, 28; current employment, 16; on Facebook discussing

temporal: scale, 34, 37, 160, 234; spatial and, 95; variation, 32, 124, 188

tene (agnate, base, source, origin, reason), 116, 124, 236

The Time of Darkness (Blong), 164

time of darkness (*taim bilong tudak; ungi*), 164–66

Tipinini (region in the eastern Porgera Valley), viii, xi, 11, 14–15, 43; bridewealth payments in, 126; census in, 181, 245n8; coffee production in, 212; cults in, 60–61, 170; household incomes in, 46–47, 245n7; migration to during drought, 170; symbols of development in, 28, 212; village councilors in, 14, 69, 122

Tisapa. *See* Yawe, Tisapa

Tiyini (clan), 142–43. *See also* landowners

Tokoyela (clan), 11; conflicts involving, 68, 113–14, 117–18, 147; *ipa ne* of, 152; marriage clusters in, 115–16; marriage patterns, 125

tombene (border, middle, in-between), 91–92, 116–17

Tongope (indigenous miner), 52, 64

totemism, 38

transformation: alchemy as, 6–7, 23, 70, 239; from development, 230–32; ontology and, 235–36; in social-ecological systems, 32–33

trees: land tenure and, 111, 179–80; species differences by elevation and forest type, 191–98. *See also* forest

Tsing, Anna: resource frontiers, 21; social forests, 107–8

Tuandaka (ritual site), 84, 133f, 135–42, 154, 165, 244n1. *See also dindi gamu*; ritual: environmental renewal

Undiki (clan), 113, 115, 117

ungi (*taim bilong tudak*; time of darkness), 164–66

University of Papua New Guinea, 15

uyenda (greed, covetousness), 146–47

van Valkenberg, Johan: disturbance effects on *Nothfagus*, 197

Vayda, Andrew: critique of political ecology, 4

village courts, 61–62, 95, 148

villages, 95–96. *See also* hamlets

Viveiros de Castro, Eduardo: richer ontologies, 39

volcanic activity, 40, 163–66, 233, 244n1

vulnerability, 14, 18, 159–61, 163, 197–99, 233, 237. *See also* resilience

Wabag, Papua New Guinea, 43, 133f; colonial violence in, 50; disease in, 59; jail in, 68–69; Mt. Hagen-Sepik Patrol in, 54, 56–57; patrols from, 62, 63; World War II impacts, 59–60

Waddell, Eric: frost in Enga, 167–68

Wage, Papua New Guinea, 90f, 138–39, 141; marriages with Porgerans, 125; migrations to Porgera, ENSO caused, 168–70

wages, 46, 72, 208, 218, 220–21

Wagner, Roy: cargo cults as reverse anthropology, 235

Wailya, Papua New Guinea, viii, 11; colonial violence in, 50–51, 53; research in, 102–3; resources in, 179

Walsh, Pat (medic for Hagen-Sepik Patrol), 52

Walters, Bradley: critique of political ecology, 4

wandakali (people), 108

wanda yama (woman spirit), 147, 150–51

wane (female descendent of clan), 116

Wangi, Kaipas, xi, 16

wapena (pandanus), 92–94

Wapena (clan), 222–23, 227

wapi (lower elevation zone), 89; characteristics of, 92, 95, 98–99, 103–4; environmental changes in, 174; land cover changes in, 184–86; location of, 177; nonoccupation of, 39; spirits of, 99–103, 150, 237; tree species in, 189–98

Wardlow, Holly: *pasindia meri*, 204; Taiyundika and Mt. Kare gold, 91

warfare: group obligations during, 120, 127; guns and, 225–26, 234; inequality and, 20; pervasiveness of, 117, 121; restraints on mobility and, 157; spirits and, 153–54; tactics, 219; wealth and, 220, 224. *See also* conflict; killings

Watiam (clan), 222

wealth: conspicuous consumption, 72; inequalities in, 2, 20, 66–67, 71, 75–76, 128,
212–15; mining and, 7, 63, 65, 226; spirits
and, 141; traditional forms of, 58–61, 94;
warfare and, 220, 234; without work, 19,
199–201, 208, 240. See also gold; money
weather. See climate; drought; frost; rainfall; temperature
West, Paige, x; conservation and development at Crater Mt., 9–10, 231
widows, 227
Wola (ethnic group), 14
work (piape), 105, 155, 200–202, 243n1; entropy and, 5, 107, 151; gender and, 108,
111–12, 207; kinship and, 78; land tenure
and, 105; money and, 200–203; in production and reproduction, 98, 109, 127,
246n1; social organization and, 118, 127;
wealth without, 19, 199–201, 208, 240
working in the life market, 2, 20, 220–21,
224–25, 227
World Bank, 199
World War II, 58, 59–60
Wuambo, Kipan, xi, 16; laita discussion,
121–23, land is ending discussion, 240;
land tenure discussion, 110; wanda yama
discussion, 150

Yakati, Muyu, xi–xii, 109; gardening with,
109–10; land claims, 222; marriage
clusters, 114; ungi discussion, 164–66;
yu koyo peya, 84

yama. See spirits
yama au (spirit attack), 146–51
yamapane (organ composed of the heart
and lungs), 146–49
yawale (Casuarina oligodon), 92, 110–12,
179–80, 189, 194–97
Yawe, Tisapa, xii; missionization discussion, 70; origin of agriculture, 83; ungi
discussion, 165–66; women and gold,
144–45
yia. See pigs
Yia One Panalia (sacred pig), 133. See also
Lemeane
yomboto (Lithocarpus sp.), 193–97
Yomondaka (clan), 121–22
Yonggom (ethnic group near Ok Tedi), 10,
231
yu. See land
yu angini wanda (earth mother women),
98–104, 150, 155, 237
yu ipane (land grease), 5, 97–98, 112, 151,
239–40. See also ipane
yu kimbuni (earth joints), 84, 136, 165–66,
240
yu koyo peya (land is ending), 3, 5, 6, 84–85,
142, 165–66, 207, 237–40
yu kulini (ground bone): land tenure rights
and, 85, 97, 239
yu umbuaini (ground skin): land tenure
rights and, 85, 97, 179, 239